ROS 2 from Scratch

Get started with ROS 2 and create robotics applications with Python and C++

Edouard Renard

ROS 2 from Scratch

Copyright © 2024 Packt Publishing

All rights reserved. No part of this book may be reproduced, stored in a retrieval system, or transmitted in any form or by any means, without the prior written permission of the publisher, except in the case of brief quotations embedded in critical articles or reviews.

Every effort has been made in the preparation of this book to ensure the accuracy of the information presented. However, the information contained in this book is sold without warranty, either express or implied. Neither the author, nor Packt Publishing or its dealers and distributors, will be held liable for any damages caused or alleged to have been caused directly or indirectly by this book.

Packt Publishing has endeavored to provide trademark information about all of the companies and products mentioned in this book by the appropriate use of capitals. However, Packt Publishing cannot guarantee the accuracy of this information.

Group Product Manager: Preet Ahuja

Publishing Product Manager: Suwarna Patil

Book Project Manager: Ashwin Dinesh Kharwa

Senior Editor: Akanksha Gupta

Technical Editor: Rajat Sharma

Copy Editor: Safis Editing

Proofreader: Akanksha Gupta

Indexer: Pratik Shirodkar

Production Designer: Jyoti Kadam

DevRel Marketing Coordinator: Rohan Dobhal

First published: November 2024

Production reference: 1291024

Published by Packt Publishing Ltd.

Grosvenor House

11 St Paul's Square

Birmingham

B3 1RB, UK

ISBN 978-1-83588-140-8

www.packtpub.com

Contributors

About the author

Edouard Renard is a software engineer, entrepreneur, and robotics teacher. In 2016, he cofounded a robotics start-up as CTO, developing and selling an educational 6-axis robotic arm based on Robot Operating System (ROS). Since 2019, he has been teaching robotics programming online. He is the best-selling instructor on Udemy for ROS 2 and has published many additional free tutorials on his website and YouTube channel (*Robotics Back-End*). He also leads offline workshops for professionals in Singapore and the US. His teaching philosophy is as follows: step by step, practical, and to the point.

I'd like to thank my friends, who supported me throughout the writing of this book, chapter after chapter. Thank you to every member of the Packt team who worked on the book—I've come to appreciate how much goes into publishing a book, beyond the writing itself. A special thanks to Lentin Joseph and Kenichi Kato for their thorough technical review, and for being great people to work with.

About the reviewers

Kenichi Kato brings over two decades of expertise in IT, manufacturing, and R&D, focusing extensively on embedded systems for autonomous and robotics applications. At the Singapore Institute of Technology, he leverages this expertise in his role, where he designs vital modules within the robotics systems curriculum that focus on electrical engineering, coding, and ROS.

His collaborative work with industry partners enhances the practical learning experience, preparing students to tackle real-world challenges with innovative solutions. Kenichi's approach not only bridges the gap between academia and industry but also equips students with essential skills for the future.

I would like to extend my heartfelt gratitude to my wife for her enduring support and patience during the writing of this review. Her encouragement was invaluable. I am also grateful to the wider robotics community, whose continuous advancements have made ROS universally accessible. Special thanks to Edouard, the author of this book, whose insightful contributions have significantly enriched our understanding and application of ROS 2.

Lentin Joseph is an author, roboticist, and robotics entrepreneur from India. He runs a robotics software company called RUNTIME Robotics in Kochi/Kerala. He is also the founder/blogger of `robocademy.com`.

He has 14 years of experience in robotics, primarily working with ROS, Gazebo OpenCV, and PCL. He has authored and reviewed more than 10 books on ROS and robotics.

He has a master's in robotics and automation from Amrita Vishwa Vidyapeetham University in India and worked at the Robotics Institute, CMU, USA. He is also a TEDx speaker.

I extend my sincere gratitude to the author, whose work I had the privilege to review. Your insights have broadened my understanding of the subject matter. I am thankful to the editorial team for their guidance and support throughout the review process. Your patience and expertise have been invaluable. Lastly, to the readers, I hope my review provides a useful perspective as you explore this book. Your engagement with my work is deeply appreciated.

Table of Contents

Preface   xiii

Part 1: Getting Started with ROS 2

1

Introduction to ROS 2 – What Is ROS 2?   3

Terminology	4	A quick story of ROS, and how we got to ROS 2	9
What is ROS, when should we use it, and why?	4	Is ROS 1 dead already?	10
Why ROS?	4	**Prerequisites for starting with ROS 2**	**10**
What is ROS?	5	Knowledge prerequisites	11
When to use ROS	8	Hardware and software	11
ROS 1 versus ROS 2	**9**	How to follow this book	12
		Summary	12

2

Installing and Setting Up ROS 2   13

Which ROS 2 distribution to choose	**14**	Installing Ubuntu 24.04 on a VM	18
What is a ROS 2 distribution?	14	**Installing ROS 2 on Ubuntu**	**25**
LTS and non-LTS distributions	15	Setting the locale	25
How to choose a ROS distribution	16	Setting up the sources	25
Installing the OS (Ubuntu)	**16**	Installing ROS 2 packages	26
The relationship between ROS 2 and Ubuntu	17	**Setting up the environment for ROS 2**	**27**
Installing Ubuntu 24.04 natively with a dual boot	17	Sourcing ROS 2 in the environment	28

Adding the source line to the .bashrc file	28	Visual Studio Code	29
		The Terminal and other tools	30
Extra tools for ROS development	**29**	**Summary**	**31**

3

Uncovering ROS 2 Core Concepts 33

Running your first node	**34**	Recap – services	47
Starting a node from the terminal with ros2 run	34	**Actions**	**47**
Introspecting the nodes with rqt_graph	35	Running an action server	48
Running a 2D robot simulation	36	A name and an interface (data type)	48
Recap – nodes	38	Sending a goal from the terminal	49
Topics	**39**	Recap – actions	50
Running a topic publisher and subscriber	39	**Parameters**	**50**
A name and an interface (data type)	39	Getting the parameters for a node	50
More experimentation with topics	41	Setting up a parameter value for a node	51
Recap – topics	43	Recap – parameters	52
Services	**43**	**Launch files**	**52**
Running a service server and client	43	Starting a launch file	52
A name and an interface (data type)	44	Recap – launch files	54
Sending a request from the terminal	45	**Summary**	**54**
More experimentation with services	45		

Part 2: Developing with ROS 2 – Python and C++

4

Writing and Building a ROS 2 Node 57

Technical requirements	**58**	What is a ROS 2 package?	61
Creating and setting up a ROS 2 workspace	**58**	Creating a Python package	62
		Creating a C++ package	63
Creating a workspace	58	Building a package	64
Building the workspace	59	How are nodes organized in a package?	65
Sourcing the workspace	60	**Creating a Python node**	**66**
Creating a package	**61**	Creating a file for the node	67

Writing a minimal ROS 2 Python node	67	Node template for Python	
Building the node	69	and C++ nodes	78
Running the node	71	Template for a Python node	78
Improving the node – timer and callback	71	Template for a C++ node	79
Creating a C++ node	73	Introspecting your nodes	80
Writing a C++ node	74	ros2 node command line	80
Building and running the node	77	Changing the node name at run time	81
		Summary	82

5

Topics – Sending and Receiving Messages between Nodes   83

What is a ROS 2 topic?	84	Introspecting topics with rqt_graph	102
A publisher and a subscriber	84	The ros2 topic command line	103
Multiple publishers and subscribers	85	Changing a topic name at runtime	105
Multiple publishers and subscribers inside one node	86	Replaying topic data with bags	106
Wrapping things up	88	Creating a custom interface for a topic	107
Writing a topic publisher	88	Using existing interfaces	108
Writing a Python publisher	89	Creating a new topic interface	110
Writing a C++ publisher	94	Topic challenge – closed-loop control	115
Writing a topic subscriber	96		
Writing a Python subscriber	97	Challenge	115
Writing a C++ subscriber	99	Solution	116
Running the Python and C++ nodes together	101	Summary	117
Additional tools to handle topics	101		

6

Services – Client/Server Interaction between Nodes   119

What is a ROS 2 service?	120	Creating a custom service interface	124
A server and a client	120	Finding an existing interface for our service	124
Multiple clients for one service	121	Creating a new service interface	126
Another service example with robotics	122	Writing a service server	128
Wrapping things up	123		

Writing a Python service server	128	Listing and introspecting services	140
Writing a C++ service server	132	Sending a service request	141
		Changing a service name at runtime	141
Writing a service client	**135**		
Writing a Python service client	135	**Service challenge – client and server**	**142**
Running the client and server nodes together	137	Challenge	142
Writing a C++ service client	138	Solution	144
Additional tools to handle services	**139**	**Summary**	**148**

7

Actions – When Services Are Not Enough — 151

What is a ROS 2 action?	**152**	Writing a Python action client	168
Why actions?	152	Creating an action client	168
How do actions work?	154	Writing a C++ action client	171
Wrapping things up	156	**Taking advantage of**	
Creating a custom action interface	**157**	**all the action mechanisms**	**173**
Defining the application and		Adding the feedback mechanism	173
the interface we need	158	Adding the cancel mechanism	176
Creating a new action interface	158	**Additional tools to handle actions**	**181**
Writing an action server	**161**	Listing and introspecting actions	182
Writing a Python action server	161	Sending a goal from the Terminal	182
Writing a C++ action server	165	Topics and services inside actions	183
Writing an action client	**168**	**Summary**	**184**

8

Parameters – Making Nodes More Dynamic — 187

What is a ROS 2 parameter?	**188**	Providing parameters at runtime	192
Why parameters?	188	Parameters with C++	194
Example of a node with parameters	188	**Storing parameters in YAML files**	**195**
ROS 2 parameters – wrapping things up	190	Loading parameters from a YAML file	195
Using parameters in your nodes	**190**	Parameters for multiple nodes	196
Declaring, getting, and using parameters		Recapping all parameters' data types	197
with Python	191		

Table of Contents | ix

Additional tools for handling parameters	198	Updating parameters with parameter callbacks	202
Getting parameters' values from the terminal	198	Python parameter callback	203
Exporting parameters into YAML	199	C++ parameter callback	204
Setting a parameter's value from the terminal	200	Parameter challenge	205
Parameter services	201	Challenge	205
		Solution	206
		Summary	208

9

Launch Files – Starting All Your Nodes at Once 209

What is a ROS 2 launch file?	210	XML versus Python for launch files	217
Why launch files?	210	Configuring nodes inside a launch file	219
Example of a launch file with seven nodes	211	Renaming nodes and communications	220
Creating and installing an XML launch file	212	Parameters in a launch file	221
Setting up a package for launch files	212	Namespaces	225
Writing an XML launch file	213	Launch file challenge	228
Installing and starting a launch file	214	Challenge	228
Creating a Python launch file – XML or Python for launch files?	215	Solution	229
Writing a Python launch file	216	Summary	232

Part 3: Creating and Simulating a Custom Robot with ROS 2

10

Discovering TFs with RViz 237

Technical requirements	238	What are TFs?	240
Visualizing a robot model in RViz	238	Links	241
Installation and setup	238	TFs	242
Starting RViz with a robot model	239	Relationship between TFs	244

Parent and child	244	**What problem are we trying to solve with TFs?**	**249**
The /tf topic	245		
Visualizing the TF tree	247	What we want to achieve	249
		How to compute TFs	250
		Summary	**251**

11

Creating a URDF for a Robot — 253

Creating a URDF with a link	**254**	**Writing a URDF for a mobile robot**	**268**
Setting up a URDF file	254	What we want to achieve	269
Creating a link	255	Adding the wheels	269
Customizing the link visual	258	Adding the caster wheel	273
		Extra link – base footprint	274
The process of assembling links and joints	**261**	**Improving the URDF with Xacro**	**276**
Step 1 – adding a second link	261	Making a URDF file compatible with Xacro	276
Step 2 – adding a joint	262	Xacro properties	276
Step 3 – fixing the joint origin	264	Xacro macros	279
Step 4 – setting up the joint type	265	Including a Xacro file in another file	280
Step 5 – fixing the visual origin	266	**Summary**	**281**
Recap – the process to follow every time	267		

12

Publishing TFs and Packaging the URDF — 283

Understanding how to publish TFs with our URDF	**284**	**Creating a package to install the URDF**	**291**
The robot_state_publisher node	284	Adding a new workspace	291
Inputs for the robot_state_publisher	285	Creating a _description package	292
Recap – how to publish TFs	286	Installing the URDF and other files	293
Starting all nodes from the terminal	**287**	**Writing a launch file to publish TFs and visualize the robot**	**295**
Publishing the TFs from the terminal	287	The XML launch file	295
Visualizing the robot model in RViz	288	The Python launch file	298
		Summary	**300**

13

Simulating a Robot in Gazebo — 303

Technical requirements	304	Spawning the robot in Gazebo	318
How Gazebo works	304	Spawning the robot from the terminal	319
Clarifying – Gazebo versus RViz	304	Spawning the robot from a launch file	320
Starting Gazebo	305	Controlling the robot in Gazebo	323
How Gazebo works with ROS 2	308	What do we need to do?	323
Adapting the URDF for Gazebo	309	Adding Gazebo systems	325
Inertial tags	310	Bridging Gazebo and ROS 2 communications	328
Collision tags	315	Testing the robot	331
		Summary	332

14

Going Further – What To Do Next — 335

ROS 2 roadmap – exploration phase	336	Learning for a specific goal	341
Common stacks and frameworks	336	What to learn for a project?	342
More exploration topics	339	What to learn to get a job?	344
		Summary	346

Index — 347

Other Books You May Enjoy — 358

Preface

In 2016, while cofounding a robotics start-up, I was searching for tools and technologies to build the software for a new six-axis robotic arm, and I stumbled upon ROS. Somehow, I got the intuition that ROS seemed to be exactly what I was looking for. Yet, I couldn't really comprehend what it was, or what it was doing.

It took me a long time to understand ROS, and as I was learning about it, I realized how painful the process was. There was—and still is—a lack of clear teaching materials and online resources, especially for beginners. As I continued to use ROS over the years, I also realized it was not just me—many other developers were still lost. This led me to create online courses to teach ROS and other related topics, with the goal of making ROS more accessible to everyone.

A few years forward, this book is a continuation of this process. With more experience using and teaching ROS (or ROS 2—the difference will be explained later), I wrote the book I wish I had when I first got started. When working on the book, I tried as much as possible to place myself in a beginner's shoes and to avoid two common obstacles to learning.

First, in tech circles, you can sometimes see toxic behavior shown by some experts who look down on you, and say stuff like "How can you not already know how to do that—it's so basic?", or explain things to you very fast while using jargon, and then make you feel stupid when you don't understand. This behavior is not helpful, and won't motivate you to learn.

Second, I don't know why, but so many people like to overcomplicate things, and this is not just related to ROS or even technology. Most of the time, once concepts are clearly understood, they can be explained in very simple steps. There's no need to make them sound complicated if it's not needed, and there's no need to spend one hour explaining something if it can be done in five minutes. That creates noise and confusion.

In this book, I want to do the opposite: teach without judgment over a lack of skills, and give priority to clear and simple explanations. With this, I hope you can learn efficiently, and finish this book being less confused than you are now, and more motivated to continue your journey with ROS 2 and robotics.

Who this book is for

This book is for engineers, researchers, students, teachers, developers, and hobbyists who want to learn ROS 2 from scratch in an efficient way, without wasting any time.

Even though you don't have to be an expert in anything, this book is not for complete beginners in software engineering. You need some good basics in Linux and Python—C++ is optional. Having a good grasp and some experience with those technologies will make your learning much easier.

No ROS (or ROS 1) experience is required.

What this book covers

Chapter 1, Introduction to ROS 2 – What is ROS 2?, explains what ROS 2 is exactly, and clears most of the doubts and confusions you could have.

Chapter 2, Installing and Setting Up ROS 2, leads you through the installation and set up of Ubuntu, ROS 2, and additional tools, so you have everything you need to work with ROS 2.

Chapter 3, Uncovering ROS 2 Core Concepts, introduces the main ROS 2 concepts through experimentation and hands-on discovery, the goal being to develop an intuition of how things work.

Chapter 4, Writing and Building a ROS 2 Node, shows you how to write ROS 2 programs, install them, and run them. Both Python and C++ are used, and additional challenges are given to make you practice more (the same applies to the following chapters).

Chapter 5, Topics – Sending and Receiving Messages between Nodes, explains how to communicate between two nodes with topics. We start with an explanation of the concept using a real-life analogy and then dive into the code.

Chapter 6, Services – Client/Server Interaction between Nodes, follows the same outline as the previous chapter—this time to work on the second most important communication type in ROS 2.

Chapter 7, Actions – When Services Are Not Enough, introduces the third and last ROS 2 communication type. This chapter is a bit more advanced and can be skipped during the first read.

Chapter 8, Parameters – Making Nodes More Dynamic, shows you how to add parameters to your nodes, in order to provide different settings at runtime.

Chapter 9, Launch Files – Starting All Your Nodes at Once, provides you with a way to start a complete ROS 2 application from a single file.

Chapter 10, Discovering TFs with RViz, introduces you to one of the most important concepts, so you can track the different coordinates of a robot over time. That will be the backbone of almost any ROS 2 application you create.

Chapter 11, Creating a URDF for a Robot, has you start a new project in which you create a custom robot with ROS 2. With URDF, you can create the robot description.

Chapter 12, Publishing TFs and Packaging the URDF, explains how to correctly package your application and generate the required TFs thanks to the URDF you have created.

Chapter 13, *Simulating a Robot in Gazebo*, teaches you how to adapt a robot for Gazebo (3D simulation tool), how to spawn the robot, and how to control it, so as to get a simulation that's as close as possible to a real robot.

Chapter 14, *Going Further – What to Do Next*, gives you more perspectives on the different paths you can take after finishing this book, depending on your personal goals.

To get the most out of this book

You need basic knowledge of the following:

- Linux, especially on how to use the command line (with auto-completion) and write code with text editors, and you should understand a bit about the file system and how the environment works (with files such as `.bashrc`).

- Python programming: most of the code will be in Python 3, using object-oriented programming. The better your Python skills, the easier it will be.

- C++ programming: you could decide to only start by following the Python examples, and thus you don't need C++. If you want to follow C++ examples as well, of course, C++ is needed.

Regarding software and operating systems, you will need to install Ubuntu on your computer (better as a dual boot, and also works with a virtual machine). This book targets Ubuntu 24.04 and ROS 2 Jazzy, but you should be able to get the most out of it with the later versions as well. Step-by-step instructions on how to install Ubuntu (in a virtual machine) and ROS 2 will be provided in the book.

If you are using the digital version of this book, we advise you to type the code yourself or access the code from the book's GitHub repository (a link is available in the next section). Doing so will help you avoid any potential errors related to the copying and pasting of code.

Download the example code files

You can download the example code files for this book from GitHub at `https://github.com/PacktPublishing/ROS-2-from-Scratch`. If there's an update to the code, it will be updated in the GitHub repository.

We also have other code bundles from our rich catalog of books and videos available at `https://github.com/PacktPublishing/`. Check them out!

Conventions used

There are a number of text conventions used throughout this book.

`Code in text`: Indicates code words in text, database table names, folder names, filenames, file extensions, pathnames, dummy URLs, user input, and Twitter handles. Here is an example: "For example, if you want to install the `abc_def` package on ROS Jazzy, then you will need to run `sudo apt install ros-jazzy-abc-def`."

A block of code is set as follows:

```
#!/usr/bin/env python3
import rclpy
from rclpy.node import Node
```

When we wish to draw your attention to a particular part of a code block, the relevant lines or items are set in bold:

```
entry_points={
    'console_scripts': [
        "test_node = my_py_pkg.my_first_node:main"
    ],
},
```

Any command-line input or output is written as follows:

```
$ sudo apt update
$ sudo apt upgrade
```

Bold: Indicates a new term, an important word, or words that you see onscreen. For instance, words in menus or dialog boxes appear in **bold**. Here is an example: "To start the VM, double click on it in **VirtualBox Manager**, or select it and click on the **Start** button."

> **Tips or important notes**
> Appear like this.

Get in touch

Feedback from our readers is always welcome.

General feedback: If you have questions about any aspect of this book, email us at `customercare@packtpub.com` and mention the book title in the subject of your message.

Errata: Although we have taken every care to ensure the accuracy of our content, mistakes do happen. If you have found a mistake in this book, we would be grateful if you would report this to us. Please visit `www.packtpub.com/support/errata` and fill in the form.

Piracy: If you come across any illegal copies of our works in any form on the internet, we would be grateful if you would provide us with the location address or website name. Please contact us at copyright@packtpub.com with a link to the material.

If you are interested in becoming an author: If there is a topic that you have expertise in and you are interested in either writing or contributing to a book, please visit authors.packtpub.com.

Share Your Thoughts

Once you've read *ROS 2 from Scratch*, we'd love to hear your thoughts! Scan the QR code below to go straight to the Amazon review page for this book and share your feedback.

https://packt.link/r/1835881416

Your review is important to us and the tech community and will help us make sure we're delivering excellent quality content.

Download a free PDF copy of this book

Thanks for purchasing this book!

Do you like to read on the go but are unable to carry your print books everywhere?

Is your eBook purchase not compatible with the device of your choice?

Don't worry, now with every Packt book you get a DRM-free PDF version of that book at no cost.

Read anywhere, any place, on any device. Search, copy, and paste code from your favorite technical books directly into your application.

The perks don't stop there, you can get exclusive access to discounts, newsletters, and great free content in your inbox daily

Follow these simple steps to get the benefits:

1. Scan the QR code or visit the link below

https://packt.link/free-ebook/9781835881408

2. Submit your proof of purchase
3. That's it! We'll send your free PDF and other benefits to your email directly

Part 1: Getting Started with ROS 2

In this first part, you will get a global overview of ROS 2 and understand what it is, when to use it, and why. You will install ROS 2, as well as all necessary tools on your computer, and discover the core concepts through hands-on experimentation.

This part contains the following chapters:

- *Chapter 1, Introduction to ROS 2 – What Is ROS 2?*
- *Chapter 2, Installing and Setting Up ROS 2*
- *Chapter 3, Uncovering ROS 2 Core Concepts*

1
Introduction to ROS 2 – What Is ROS 2?

Robot Operating System (**ROS**) can be confusing, as evidenced by its name. It's difficult to know what it is exactly, what it contains, and what it does. Also, why do you even need ROS, and when should you use it?

Before getting started, it is okay to be confused—most people are. Although ROS is one of the best tools to learn and develop robotics applications, it also comes with a steep learning curve, with the first roadblock being understanding what it is.

In this quick first chapter, I will explain the terminology we will use throughout this book. You will then see why ROS exists, and what problems it can solve for you. After that, we will dive a bit deeper into the four pillars of ROS to understand what it is. You will also see a few examples of when and when not to use it.

By the end of this chapter, you will have a better understanding of the global picture behind ROS and be clear of the most common confusions. You will also understand what prerequisites you need before you get started with ROS, as well as how to follow this book to get the most out of it. This will help you get started on the right foot.

In this chapter, we are going to cover the following topics:

- Terminology
- What is ROS, when should we use it, and why?
- ROS 1 versus ROS 2
- Prerequisites for starting with ROS 2
- How to follow this book

Terminology

You might have seen the terms ROS, ROS 1, ROS 2, and other kinds of variations (with or without a space), which can be confusing.

Let's clear this up now:

- ROS 1 is (was) the first version of ROS
- ROS 2 is the second and newer version of ROS and will be the focus of this book

In this book, I will use the following convention:

- **ROS**: When talking about general ROS concepts, philosophy, and so on
- **ROS 1**: When talking specifically about the first version of ROS. However, this will be quite rare since the focus here is on ROS 2
- **ROS 2**: When talking about the second version of ROS

> **Note**
> I may sometimes write *ROS* or *ROS 2* interchangeably since we won't be focusing on ROS 1 here.
>
> It's not impossible that, in the future (when ROS 1 has completely disappeared), the name ROS 2 becomes ROS again. If you've heard about Angular, it started as AngularJS, after which they released Angular2, and then a few years later, it simply became Angular. I guess that something similar will happen with ROS, although this is only a theory of mine for now.

What is ROS, when should we use it, and why?

Before we start understanding what ROS is, let's understand why we would need it.

Why ROS?

Let's start with a big problem that occurs often in robotics.

Imagine that you just got a new project at your job, and you have to develop a robotics application, or you are doing a new research thesis. One very important thing to take into account is that in real life, any project or thesis will have a specific duration, from a few months to a few years.

Now, what will happen?

You start to design the robotics system you need for your project and soon realize that it will take a lot of time to develop the robot because all the existing solutions you found don't match what you need. After a few weeks, you finalize the specifications, and you start building your robot. A few months in, you're still developing the basic software for wheel control and navigation. You underestimated how much time it would take you to just get the robot running. After 1 or 2 years, you realize that all you've done for now is build a robotics system, and you still haven't started the core functionality of your application or research. Now is the time to hurry.

You finish the robot as well as you can, make some shortcuts, and publish your paper or present that prototype. In the best-case scenario, you could also share your code with an open-source license so that other people can use it, but probably not directly as it's just code for your own need, not a complete framework or library with modular components, documentation, and so on.

Then, you move on to a new project, new job, and new research. Somebody else will take your place, read your code, and realize that it doesn't help them build their application. Hence, they have to start from scratch.

What just happened here is that you reinvented the wheel. The next person will repeat the same circle. And this is much more common than you think. People keep reinventing the wheel over and over again. This is the number one reason why ROS was created: to stop you from reinventing the wheel anytime you need to create a robot. Just like you have open-source frameworks, tools, and environments to develop websites or mobile applications, why not do the same for robotics? This is the philosophy behind ROS: to provide a *standard* for robotics applications that you can use on *any robot*.

After you learn ROS, you can spend less time on the basics and focus on the key functionalities you want to add instead. You can program new robots in no time, join existing projects, and easily collaborate with a team.

What is ROS?

ROS is hard to define because it's not just one thing. And to be honest, I don't think you can truly understand what it is until you start to understand how to write code with it.

One thing we can start with is what ROS is not.

ROS is not an operating system. It's a combination of four main parts:

- Framework
- Set of tools
- Plug-and-play plugins
- An online community

Let's dive a bit more into each of these parts.

ROS is a framework with plumbing

ROS comes with a set of rules on how to build an application. As we will see in this book, you will need to create packages, and then write programs inside those packages (nodes). There is a specific way to create and write them, as well as create tools to build and use them.

Any framework comes with a specific set of rules. The remarkable thing about this is that after you've created a few projects, any new project is going to be easier and faster to set up. Also, as everyone is following the same set of rules, you can more easily work in a team or understand and use the code written by others.

As a direct consequence of using this framework, you get access to what is often referred to as **plumbing**, which means that the underlying communication between the nodes is managed for you. Imagine that you're building a house, and the plumbing or electrical system is already done for you. This will save you a lot of development time, and you also don't need to learn how to do it yourself (and thus, reinvent the wheel).

To sum it up, with ROS, you can easily separate your application into different sub-programs (called **nodes**). The communication between nodes is handled for you. You can easily test one component, and if this component fails, it will not affect the other running components. ROS is a modular framework.

A set of tools

ROS comes with a set of tools that allow you to develop faster. Among them, you can find command-line tools to build the application, introspection tools to monitor the flow of communication, logging functionalities, plots, and more.

You also get 3D visualization tools to see what your robot is doing, and even a complete simulator using real physics, named **Gazebo**, so that you can work on a realistic simulation before trying out your robot.

There are quite a lot of available tools, and we'll discover many of them throughout this book. As an example of how useful it can be, there is one (called **bags**) that allows you to save communication streams so that they can be replayed later. Let's say you build a mobile robot, and you need to test the robot outside when it's raining, then continue to develop the software while taking the rain into account. You probably won't have rain every day, or you won't even have access to the robot any time you want. With this tool, you can run the experiment once, save the data, and replay it later to develop your application for a specific set of conditions.

Capabilities – plug-and-play plugins and stacks

This is probably where you will save hundreds of hours. Imagine two common scenarios:

- You develop a mobile robot and need the robot to navigate autonomously in a dynamic environment.
- You develop a six-axis robotic arm and want to create motion planning to perform a smooth movement on all axes.

This looks quite complex and involves understanding and implementing several algorithms, as well as writing well-optimized and efficient code. This is where you would probably have to reinvent the wheel and waste lots of precious time.

For those two scenarios, you can find existing plugins that do the job for you. All you need to do is install the plugins and configure your robot to make it compatible. Of course, this is easier said than done, but the workload can be counted in days/weeks, not months/years. And once again, once you know how to use those plugins, your next project will take much less time.

There are many plugins that you can use. Some are quite simple, while others involve a collection of plugins and are also called frameworks or stacks. Your job as a ROS developer is to *glue* all those components together, and maybe create new components for functionalities that are not developed yet.

Online community

This is the fourth pillar of ROS, and it's quite an important one: the community. ROS is an open-source project with a permissive license. I can't give you any legal advice on licensing, but you can use ROS in a commercial product without having to redistribute your code.

You can find all the ROS code online, as well as the code for the plug-and-play plugins. Everything is easily accessible on GitHub.

The ROS project is also backed by an online community that you can most commonly find in the following areas:

- **Robotics Stack Exchange** (`https://robotics.stackexchange.com/`): You can use this to ask technical questions. If you know Stack Overflow, as most developers do, well, this is Stack Overflow for robotics.
- **ROS Discourse forums** (`https://discourse.ros.org/`): Here, you can get informed about the latest developments, jobs, community projects, new ideas, and more. I recommend checking this website often to stay up to date with where ROS is going.

When to use ROS

Now that you understand a bit more what ROS is, should you use ROS whenever your project has something 'robotics' in it? In this section, I will give you some hints on when using ROS makes sense, backed by some examples to give you a better idea.

First, if you're reading this book because you need to learn ROS for your work/university, then the question is easily answered: yes, you will use ROS for your project.

But if you must make the decision yourself, what should you do?

Let's simplify robotics and say that a robotics system contains three categories of things: actuators, sensors, and controllers.

An **actuator** is something that creates movement (for example, a motor to rotate a wheel). A **sensor** will read data from the environment (for example, a camera, laser scan, or temperature sensor). A **controller** is something that is in between: it takes the data available from one or multiple sensors (input) and, through an algorithm, creates a command for the actuators of the robot (output). In a way, the controller is the *brain* or one of the brains of the robot.

For very simple applications, when you just have a few sensors and actuators, you might not need ROS.

Here are a few examples where ROS isn't needed:

- You just need to take a picture from a camera when a user presses a button, using a Raspberry Pi board, and send this picture to a web server. There's no need to use ROS—you can just combine a few Python libraries in a script, and you're done. Using ROS here would be a good example of over-engineering (unless you do this for learning purposes).

- You have to use a servo motor to open/close a door when a movement is detected, using an infrared sensor. This is a very simple application that can easily be programmed using a basic microcontroller board—and you can do a quick prototype with a board such as Arduino.

- You have built a simple robot with two wheels and an infrared sensor, and you want to make the robot follow a line. This is a typical project that's given to students in engineering school, and a simple algorithm on an Arduino board will do.

Now, let's consider some examples where ROS is needed:

- You have a new mobile robot with two wheels and a laser scan, and you want to read data from the laser scan, map the environment, make the robot move autonomously, and control the two wheels accordingly. On top of that, you want to simulate the robot in 3D with real physical properties. This is when ROS is going to become very handy. Not only will it help you to make all the components work together, but you can also use existing algorithms for path planning (through a ROS plugin) and simulation.

- You need to create a system that contains a six-axis robotic arm, or even multiple robotic arms working together, along with conveyor belts and mobile robots.
- Your robotics application (not necessarily just one robot) contains lots of sensors and actuators that you want to develop separately and add them in a modular way.
- You want to create a hardware driver for a component and make this component easy to use by other robotics developers. By making the component *ROS-compatible*, anybody who knows ROS can integrate it into their application with low effort.

As you can see from the former examples, ROS is not always needed whenever you need to program hardware or create a robotics system. Of course, you could use it for any application, but it's like if you were to use a complete web framework (for example, Django) for a single static web page.

With the latter examples, you can see that if your system becomes more complex, if you want to easily collaborate with other developers, or if you realize that one big part of your system can be solved with one of the plug-and-play plugins, ROS may be the solution.

Of course, it takes time to learn it and your first project will take longer to complete, but then, with more experience, you will go much faster.

As an example, it could take less than a week for a senior ROS developer to write custom code for a robotic arm (including robot model, motion planning, and hardware control) and the same for a mobile robot with navigation capabilities (provided that the hardware already has a ROS driver). Less than a week and you get a working software prototype.

ROS 1 versus ROS 2

To be clear, this book is all about ROS 2, not ROS 1. You will start learning ROS 2 from zero experience. This section is probably the only time I will be talking that much about ROS 1.

A quick story of ROS, and how we got to ROS 2

ROS 1 (originally called ROS) was first developed in 2007. It quickly gained popularity and grew exponentially in the following years.

In 2014, the ROS 2 project was announced. Simply put, ROS 1 was a bit too limited for industrial applications (lack of real-time support, safety, and so on) and was only used in research/education. To solve this problem, the developers decided to make ROS more "industrial friendly," as well as make it better, thanks to all the lessons learned from the beginning of ROS.

Now, why create ROS 2 and not just continue ROS with some new changes? Well, the changes were too big, and they would have completely broken compatibility with older versions. Thus, it was decided to create a completely new ROS from scratch and name it ROS 2. In 2014, ROS 2 was officially announced, and the development of the project started.

In December 2017, the first ROS 2 distribution was released, which meant that ROS 1 and ROS 2 started to co-exist. At this point, ROS 2 was lacking many core functionalities and plugins, making it unsuitable for serious projects. Most ROS developers were still using ROS 1.

Years passed by and ROS 2 got more development, plugins, and more. Its popularity started to grow.

I would say it was worth it to use ROS 2 (compared to ROS 1) starting from 2022. This is probably more of a personal opinion and some people might disagree, but from 2022 and the release of **ROS 2 Humble** (more on distributions in *Chapter 2*), we had access to a long-term release that was stable, with all the major plugins and stacks working correctly, which is what you need to program a robot.

In the meantime, it was announced that ROS 1 would end in May 2025. After this date, ROS 1 would still exist, but it wouldn't be supported anymore.

2023 was the year with the most significant shift from ROS 1 to ROS 2 among the ROS community. It is now safe to say that ROS 2 is the way to go when developing new ROS applications.

So, if you had previously heard about ROS 1 and ROS 2, now you know that ROS 2 is what you need to learn, and we can say that ROS 1 is a dead project. But is that true?

Is ROS 1 dead already?

In theory, yes, but in practice, it's (always) a bit different. As you probably know, several companies are using obsolete and legacy technologies. The reason is that updating software to a new version is often quite expensive and can also be risky. That's why you still see job offers from banking systems requiring skills in Cobol, a programming language from the 1960s that no one uses anymore.

In robotics, things are a bit similar. Some companies have released robots with a specific version of ROS 1, and while the robot is still on the market, the company will not upgrade and still use and maintain the previous version, also called *legacy*. Thus, the definitive transition in 2025 is going to take a few more years.

Why am I writing this? Simply to let you know that if you happen to get a job in a robotics company that has been using ROS already, you might encounter a few ROS 1 projects, even after ROS 1 is officially finished. However, be assured that all the ROS 2 knowledge you have can easily be ported to ROS 1 as the core concepts are the same.

To conclude, for all new learnings, projects, studies, teaching, and startups, ROS 2 is what you need. I will now close this chapter of ROS 1 and focus on ROS 2. As mentioned previously, I might write *ROS* or *ROS 2* interchangeably as we aren't targeting ROS 1 here.

Prerequisites for starting with ROS 2

To get started with ROS and this book, there are a few things you need to know.

Knowledge prerequisites

It is best that you have some knowledge of the following:

- **Linux command line**: Since we'll be using **Ubuntu**, being familiar with Linux is mandatory. You don't need to be an expert—you just need to know the basics. Many tools in ROS 2 involve the command line, so knowing how to open a Terminal and write basic commands will help you tremendously.
- **Python programming**: The two most common languages for ROS are **Python** and C++. Python is easier to get started with and allows you to prototype things faster. Hence, this is the language we will use for all detailed explanations. You need to know Python basics, and **object-oriented programming** (**OOP**) is a good plus as ROS 2 is heavily using OOP everywhere.
- **Optional**: C++ programming. Even if the focus of the book is on Python, I still wanted to include C++ code for everything we do. If you only want to learn Python, you can ignore the C++ code, but of course, if you want to follow C++ instructions, you need C++ basics (better with OOP).

I want to emphasize that it will be much, much easier for you to learn ROS 2 if you have good programming and Linux basics. Learning ROS is already quite challenging (though with this book, the goal is to reduce the learning curve), so if you're starting ROS, Linux, and Python from scratch, this could be overwhelming.

If you're reading these lines and you don't know how to write a Python function or navigate to a directory in the Terminal, then I really recommend that you pause here, take some time to learn Python and Linux basics, and come back to this book. There's no need to spend hundreds of hours doing this but investing some time to get the basics right will help you finish this ROS 2 book faster.

Hardware and software

You'll need to have a computer to follow this book. Regarding the specifications, you don't need anything fancy to get started with ROS 2. If you can open a web browser with a few tabs and have a smooth experience, I would say that your computer is good enough to get started.

Then, later, depending on what you want to do with ROS, you might need a better machine (for example, if you want to simulate multiple robots using lots of sensor and image processing). However, it's probably better to wait until you need the extra power to upgrade. For now, the most important thing to do is start learning ROS.

For software requirements, I will give you the necessary installation instructions throughout this book. All the software we will be using is free to use and open source.

We will also use Ubuntu 24.04, in which we will run ROS 2. Having Ubuntu installed is a requirement, but I will give you a recap in *Chapter 2*.

How to follow this book

The book is divided into three parts, including 14 chapters.

Each chapter can be followed individually, although for one chapter, you need the knowledge from all previous chapters.

If you got this book because you just want to get started from scratch, then it's simple: follow the book in the order it's been written. I have designed it specifically so that you learn the concepts one step at a time without having to think about what directions you should take.

Then, as you progress, feel free to come back to any chapter to clear up doubts. I encourage you to do that. The first time you learn about a concept, you don't necessarily grasp all the subtleties. As you continue with this book and use the concept along with other new concepts, you often have 'epiphany moments', when everything clicks together.

If you already know some ROS 2 basics (or you've already read this book), then feel free to jump to any chapter. If a chapter starts from a code base that we developed in previous chapters, then you will be able to download the code and start from there.

There is a GitHub repository you can use to follow this book: `https://github.com/PacktPublishing/ROS-2-from-Scratch`. All the code we'll write is hosted there, so be sure to use this GitHub repository closely while following along. I will explain how to use this repository a bit later in this book.

Summary

In this introductory chapter, we cleared up some of the most common confusion points regarding ROS: its name, what it is and isn't, when to use it, and why. You also learned more about the different ROS versions (ROS 1 and ROS 2), and you learned what kind of prerequisites you need to get started with ROS 2.

You should now have a better understanding of the big picture, and even if everything still seems a bit confusing, don't worry too much—it will all make sense when you use the ROS 2 concepts and code with them.

Now, to be able to use ROS 2, we need to install it. This will be the focus of the next chapter and will help you get your environment 100% ready for ROS 2.

2
Installing and Setting Up ROS 2

Before using ROS 2, we need to install it and set it up. Doing this is not as trivial as just downloading and installing a basic program. There are several ROS 2 versions (called **distributions**), and we need to choose which one is the most appropriate. We also need to pick an Ubuntu version as ROS and Ubuntu distributions are closely linked.

Once you know which ROS/Ubuntu combination you need, you have to install the corresponding Ubuntu **operating system** (**OS**). Although being familiar with Linux is a prerequisite for this book, I will still do a recap on how to install Ubuntu on a **virtual machine** (**VM**), just in case, so you won't be lost and can continue with this book.

Then, we will install ROS 2 on Ubuntu, set it up in our environment, and install additional tools that will allow you to have a better development experience.

By the end of this chapter, you will have everything ready on your computer so that you can use ROS 2 and write custom programs.

Even if all the installation steps sound a bit daunting, don't worry—it's not that hard, and it gets easier to do with every new installation. To give you an idea, with a stable internet connection, it takes me about 1 hour to install a fresh version of Ubuntu and 20 minutes to install ROS (most of that time is spent waiting for the installation to finish).

In this chapter, we are going to cover the following topics:

- Which ROS 2 distribution to choose
- Installing the OS (Ubuntu)
- Installing ROS 2
- Setting up the environment for ROS 2
- Extra tools for ROS 2 development

Which ROS 2 distribution to choose

Before you install ROS 2, it's important to know which distribution you need to use. To make that decision, you first need to understand a bit more about what ROS 2 distributions are, and what specificities each one has.

What is a ROS 2 distribution?

ROS 2 is a project in continuous development, constantly receiving new features or improvements to existing ones.

A distribution is simply a *freeze* in the development at some given point to create a stable release. With this, you can be sure that the core packages for one given distribution will not have any breaking changes. Without distributions, it would be impossible to have a stable system, and you would need to update your code constantly.

Every year, a new ROS 2 distribution is released on May 23. This day corresponds to *World Turtle Day*. As you will be able to observe, all ROS distributions have a turtle as a logo; there's a mobile robot platform named **TurtleBot** and even a 2D educational tool named **Turtlesim**. This is based on a reference to an educational programming language from 1967 named *Logo*, which included a feature to move some kind of *turtle robot* on the screen. So, if you were confused about why there are so many turtles everywhere, now you know—and that's the end of this turtle parenthesis.

You can see all ROS 2 distributions on the ROS 2 documentation releases page: `https://docs.ros.org/en/rolling/Releases.html`.

You will see one new distribution every year in May. As for the order, there is no number; instead, the names are in alphabetical order. The first official release was named *Ardent Apalone*, then *Bouncy Bolson*, and so on. In May 2024, **ROS Jazzy Jalisco** was released. Following this, you can expect to have *ROS K* in 2025, *ROS L* in 2026, and so on. The name of a new release is usually announced 1 year in advance.

> **Note**
>
> ROS distributions contain two names, but it's common practice just to refer to the first one. So, instead of talking about ROS Jazzy Jalisco, we will talk about **ROS Jazzy**. We could also write *ROS 2 Jazzy* to specify that this distribution is for ROS 2, not ROS 1, but this isn't needed since Jazzy is a name only used for ROS 2, hence ROS Jazzy.

On top of all the displayed distributions, there is another one that exists in parallel: **ROS Rolling**. This distribution is a bit special and is the distribution where all new developments are made. To make an analogy with Git and versioned systems, it's just like having a *development* branch and using this branch to release stable versions once a year. Thus, ROS Rolling is not a stable distribution, and I don't recommend using it for learning or to release a product. This is a distribution you can use if you want to test brand-new features before they are officially released into the next stable distribution—or if you want to contribute to the ROS code. However, if you're reading this book, you're not there yet.

Now that you know what ROS 2 distributions are and how to find them, let's start to look at the differences between them. This will allow us to choose the right one.

LTS and non-LTS distributions

If you look a bit closer, you'll see that some distributions are supported for 5 years, others for 1.5 years (this only applies after 2022). You can see this by comparing the release date with the **end-of-life** (**EOL**) date. Currently supported distributions also have a green background on the screen, so you can easily spot them.

When a distribution reaches its EOL date, it just means that it will not receive official support and package updates anymore. This doesn't mean you can't use it (in fact, lots of companies are still using legacy versions from 5 years ago or more), but you won't get any updates.

The first official ROS 2 distribution was *ROS Ardent*, released in December 2017. After that, the first few distributions were still not quite complete, and the development team preferred to release shorter distributions so that the development could go faster.

ROS Humble was the first **Long-Term Support** (**LTS**) release supported for 5 years (2022-2027).

ROS Jazzy is also an LTS version, with official support from 2024 to 2029. From this, you can expect that every 2 years (even number: 2024, 2026, 2028, and so on), a new LTS distribution will be released in May and supported for 5 years.

A few LTS distributions can coexist. In 2026, for example, with the release of the *ROS L* distribution, you will be able to use ROS Humble and ROS Jazzy as well.

Then, you have non-LTS distributions. Those are released on odd years (2023, 2025, 2027, and so on) and are supported for 1.5 years only. Those distributions are released just so you can have access to new development in a somehow stable release without having to wait for 2 years. However, due to the short lifespan of non-LTS distributions and the fact that they will probably be less stable (and less supported), it is best not to use them if your goal is learning, teaching, or using ROS for a commercial application.

With this, you can see that we can discard half of the distributions and now focus only on the LTS ones that are currently supported. Let's finish this section and choose the distribution that we will use for this book.

How to choose a ROS distribution

What I recommend is to use the latest available LTS distribution. However, I wouldn't necessarily use an LTS distribution just after it's been released because it can still contain some bugs and problems. Also, some of the plugins and other community packages you need might not have been ported yet. Generally, if you want to work with a stable system, it's sometimes best not to stay too close to the new and shiny technology and wait a bit.

For example, ROS Humble was released in May 2022. Right after it was available, I tested it, but to use it in a production environment, I would have had to wait until September or even November, just to be sure everything was working correctly.

So, for this book, we will use ROS Jazzy, which was released in May 2024.

> **Note**
> You can learn with one distribution and then start a project with another one. If you have a project or job that requires you to use a different ROS 2 distribution, you can still start to learn with ROS Jazzy. The gap between distributions is very small, especially for the core functionalities. 99% of this book can be applied to any ROS 2 LTS distribution released after 2022.

Installing the OS (Ubuntu)

ROS 2 works on three OSs: Ubuntu, Windows, and macOS. Although Ubuntu and Windows get Tier 1 support, macOS has only Tier 3 support, meaning "best effort," not "fully tested." You can learn more about what Tier 1, 2, and 3 mean on the REP 2000, which describes the timeline and target platforms for ROS 2 releases: `https://www.ros.org/reps/rep-2000.html`.

This means that using macOS for ROS 2 is not necessarily the best choice for learning (if you're an Apple user). We're left with Windows or Ubuntu.

From teaching experience, I saw that even if ROS could work well on Windows, it's not easy to install and use it correctly. Lots of bugs can occur, especially with the 2D and 3D tools. When you're learning ROS, you want a smooth experience, and you want to spend time learning the features, not fixing the configuration.

Hence, the best overall option is to use Ubuntu. If you don't have Ubuntu and are using Windows/macOS, you can either install Ubuntu natively as a dual boot on your computer or use a VM (there are a few other options, but I won't cover those here).

Now that we have selected ROS Jazzy, and we want to run it on Ubuntu, the question is: on which Ubuntu distribution do we install it?

The relationship between ROS 2 and Ubuntu

If you go to the Jazzy release page (`https://docs.ros.org/en/rolling/Releases/Release-Jazzy-Jalisco.html`), you'll see that ROS Jazzy is supported on **Ubuntu 24.04** (and not any other previous or future Ubuntu distributions).

There is a close relationship between ROS and Ubuntu distributions. This relationship is quite simple: for every new Ubuntu LTS distribution (every 2 years on an even number), there is a new ROS 2 LTS distribution:

- **Ubuntu 22.04**: ROS Humble
- **Ubuntu 24.04**: ROS Jazzy
- **Ubuntu 26.04**: ROS L

It's important to use the correct combination. Thus, before installing ROS Jazzy, the first thing you must do is make sure you have Ubuntu 24.04 installed on your computer. If you happen to have an older version, I strongly encourage you to upgrade or simply install Ubuntu 24.04 from scratch.

> **Note**
>
> If you have to use another Ubuntu distribution because, for example, you're using a computer from school/work and you can't change the OS, then use the corresponding ROS distribution. However, I recommend **not** using anything older than ROS Humble and avoiding non-LTS distributions. You could also install Ubuntu 24.04 on a VM (as described a bit later).

You have probably already installed a Linux OS at some point in your life, but, from experience, I know some people reading this could get a bit lost with the installation. Hence, I will provide additional installation instructions—an overview of dual boot and detailed instructions for VMs. Feel free to skip this and go to the ROS 2 installation section if you already have installed Ubuntu.

Installing Ubuntu 24.04 natively with a dual boot

The best option is to have Ubuntu installed natively on your computer. This will allow you to follow this book and then go further without any problems. I won't provide a complete tutorial on how to do that here; you can easily find lots of free tutorials on the internet.

Here are the high-level important steps you have to follow:

1. Free some space on your disk so that you can create a new partition. I recommend a minimum of 70 GB, more if possible.
2. Download an Ubuntu `.iso` file from the official Ubuntu website (Ubuntu 24.04 LTS).
3. Flash this image on an SD card or USB key (with a tool such as Balena Etcher).

4. Reboot your computer and choose to boot from the external device.
5. Follow the installation instructions. **Important**: When asked how you want to install, select **Alongside Windows**, for example—don't erase all your disk.
6. Complete the installation. Now, when you boot your computer, you should get a menu where you can select whether you want to start Ubuntu or Windows.

Those are the main steps to follow; you can find all the information you need on the internet.

Installing Ubuntu 24.04 on a VM

If you can't install Ubuntu as a dual boot (due to technical restrictions, lack of admin rights on the computer, or something else), or if you want to get started quickly with not too much effort, just for the sake of learning ROS, then you might wish to use a VM.

A VM is quite easy to set up and is useful for teaching and learning purposes. As an example, when I teach ROS offline in a workshop for beginners, I often provide a VM that contains everything already installed. With that, it's easier for the participants to get started quickly with ROS. Later, when they have more knowledge, they can take the time to set up a proper OS by themselves.

> **Note**
> *Part 3* of this book (regarding 3D simulation and Gazebo) will probably not work well on a VM. You can still use a VM for most of this book and set up a dual boot at the end.

I will now show you how to install Ubuntu 24.04 on a VM so that you can do it even with no prior knowledge of how to create and run a VM.

Step 1 – downloading the Ubuntu .iso file

Download the Ubuntu 24.04 `.iso` file `https://releases.ubuntu.com/noble/`. Note that, just like ROS, Ubuntu distributions also have a name. For Ubuntu 24.04, the name is *Ubuntu Noble Numbat*. We usually only use the first name, so **Ubuntu Noble** in this case.

Click on **64-bit PC (AMD64) desktop image**. The file should be 5 to 6 GB, so make sure you have a good internet connection before downloading it.

Step 2 – installing VirtualBox

You can start *Step 2* while the Ubuntu `.iso` file is being downloaded.

Two popular VM managers have a free version: VMware Workstation and VirtualBox. Both would work, but here, I will focus on VirtualBox as it's slightly easier to use.

Go to the download page of the official VirtualBox website: https://www.virtualbox.org/wiki/Downloads. Under **VirtualBox Platform Packages**, select the current OS you are running. If you want to install VirtualBox on Windows, for example, choose **Windows hosts**.

Download the installer, then install VirtualBox like any other software.

Step 3 – creating a new VM

Once you've installed VirtualBox and downloaded the Ubuntu .iso file, open the VirtualBox software (VirtualBox Manager) and click **New**. This will open a pop-up window where you can start to configure the new VM:

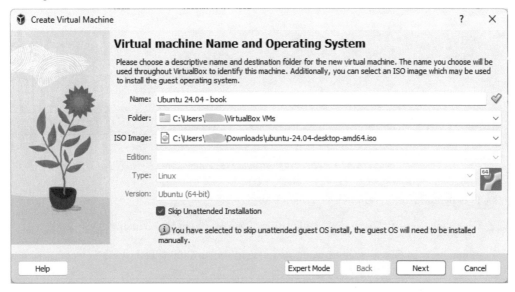

Figure 2.1 – Starting the VM setup process

Here, we have the following values:

- **Name**: Name the machine. This can be anything you want, just so you can recognize the machine—for example, Ubuntu 24.04 - book.
- **Folder**: By default, VirtualBox will create a VirtualBox VMs folder in your user directory, where it will install all the VMs. You can keep this or change it if you want.
- **ISO Image**: This is where you select the Ubuntu .iso file that you've just downloaded.
- **Type**: This should be Linux.
- **Version**: This should be Ubuntu (64-bit).
- **Skip Unattended Installation**: Make sure you check this box. Leaving this unchecked could be a cause of issues later.

Click **Next**. Here, you'll need to choose how much CPU and RAM you want to allocate for the machine:

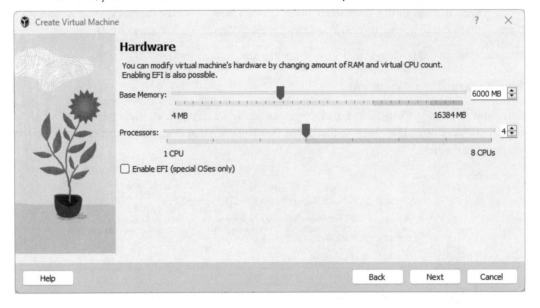

Figure 2.2 – Allocating hardware resources for the VM

This will depend on your computer's configuration.

Here's what I recommend for the RAM allocation (**Base Memory** on VirtualBox):

- If you have 16 GB or more on your computer, allocate 6 GB (like I did in *Figure 2.2*) or a bit more; this is going to be enough.
- If you have 8 GB RAM, allocate 4 GB.
- For less than 8 GB, adjust the RAM value (you can set a value now and modify it later in the settings) so that you can start the VM, open **VS Code** and Firefox with a few tabs, and your machine doesn't slow down too much. If things get too slow, consider using a more powerful machine for learning ROS.

For the CPU allocation, allocate half of your CPUs. So, if your computer has 8 CPUs, set the value to 4. In my setup, I have 4 CPUs and 8 logical processors, so I chose 4 CPUs. Try not to go under 2 as it will be very slow with only 1 CPU. If in doubt, try one setting now; you can change it later.

All in all, it's better to stay in the green zone. If you have to push to the orange zone, then make sure that when you run your VM, you don't run anything else on your computer (or maybe just a web browser with one tab or a PDF reader for this book).

Click **Next**. The last thing to do for now is to configure the virtual hard disk that will be created for the VM:

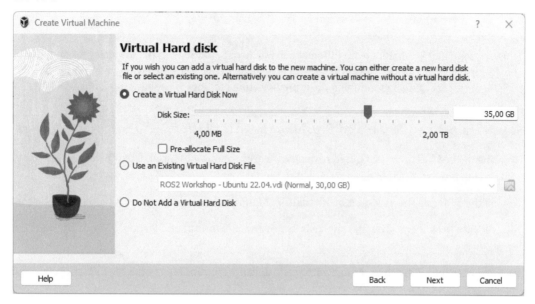

Figure 2.3 – Creating a virtual disk for the VM

Here are the settings you must choose for this screen:

1. You can keep the default option (**Create a Virtual Hard Disk Now**). The default size is 25 GB. To learn ROS 2, I recommend using 30 to 40 GB minimum. Anyway, the size of the VM will start low and expand as you install more things, so you can set a higher maximum without blocking resources.
2. Keep the **Pre-allocate Full Size** box unchecked.

Click **Next**. You will now get a recap of all options you chose in the previous steps. Then, click **Finish**. You will see the newly created VM on the left in VirtualBox Manager.

Before you start the VM, there are a few more things we need to configure. Open the settings for the VM (either select it and click on the **Settings** button or right-click on the VM and choose **Settings**).

Modify the following three settings:

- **System | Acceleration**: Uncheck the **Enable Nested Paging** box
- **Display**: Uncheck the **Enable 3D Acceleration** box (this one may already be unchecked)
- **Display**: Increase **Video Memory** to 128 MB if possible

With these settings, you can probably avoid unexpected behaviors and problems with the graphical interface in the VM.

> **Note**
>
> Every computer is different, with different hardware configurations. What works for me and lots of people might not work the same for you. If you experience weird behavior when running the VM, maybe try again by modifying those previous three settings. Test only one change at a time.

Step 4 – starting the VM and finishing the installation

Now that the VM has been configured correctly, we need to start it to install Ubuntu, using the Ubuntu `.iso` file that we've downloaded and added to the settings.

To start the VM, double-click on it in **VirtualBox Manager**, or select it and click on the **Start** button.

You will get a boot menu. The first choice is **Try or Install Ubuntu** and it should already be selected. Press *Enter*.

Wait a few seconds; Ubuntu will start with the installation screen. Follow the configuration through the different windows:

1. Choose your language. I recommend you keep **English** so you have the same configuration as me.
2. Skip the **Accessibility** menu, unless you need to set up a bigger font size, for example.
3. Select your keyboard layout.
4. Connect to the internet. To do so, choose **Use wired connection**.
5. At this point, you might have a screen asking you to update the installer. In this case, click **Update now**. Once finished, click **Close installer**. Look for **Install Ubuntu 24.04 LTS** on the VM desktop and double-click on it. This will start the installation again from *Step 1*; repeat *Steps 1 to 4*.
6. When you're asked how you would like to install Ubuntu, choose **Interactive installation**.
7. For the applications to install with Ubuntu, go with **Default selection**. This will reduce the space used by the VM and you still get a web browser, as well as all the core basics—we don't need more than that to install ROS 2.
8. When asked if you want to install recommended proprietary software, choose **Install third-party software for graphics and Wi-Fi hardware**.
9. For the disk setup, select **Erase disk and install Ubuntu**. There's no risk here as we're *erasing* the empty virtual disk we've just created (if you were installing Ubuntu as a dual boot, you would have to choose another option).

10. Choose a username, computer name, and password. Keep things simple here. For example, I use `ed` for the user and `ed-vm` for the computer's name. Also, make sure that the password is typed correctly, especially if you've changed the keyboard layout previously.
11. Select your time zone.
12. On the recap menu, click **Install** and wait a few minutes.
13. When the installation is complete, a popup will ask you to restart. Click **Restart now**.
14. You will see a message stating **Please remove the installation medium, then press ENTER**. There's no need to do anything here—just press *Enter*.
15. After booting, you will get another welcome screen popup. You can skip all the steps and click **Finish** to exit the popup.

With that, Ubuntu has been installed.

Just to finish things properly, open a Terminal window and upgrade the existing packages (it's not because you just installed Ubuntu that all packages are up to date):

```
$ sudo apt update
$ sudo apt upgrade
```

That's it for the installation. However, there's one more thing specific to VirtualBox that we must do so that the VM window works correctly.

Step 5 – Guest Additions CD Image

At this point, if you try to resize the window where the VM is running, you will see that the desktop resolution doesn't change. Also, if you try to copy/paste some text or code between your host and the VM, it probably won't work.

To fix this, we need to install what's called the *Guest Additions CD Image*.

First, open a Terminal and run the following command (note that copy/paste doesn't work yet, so you have to type it manually). This will install a few dependencies, all of which are required for the next step:

```
$ sudo apt install build-essential gcc make perl dkms
```

Then, on the top menu of the VM window, click **Devices | Insert Guest Additions CD image**.

You will see a new CD image on the Ubuntu menu (on the left). Click on it—this will open a file manager. Inside the file manager, right-click on the empty space and choose **Open in Terminal**. See the following figure for more clarification:

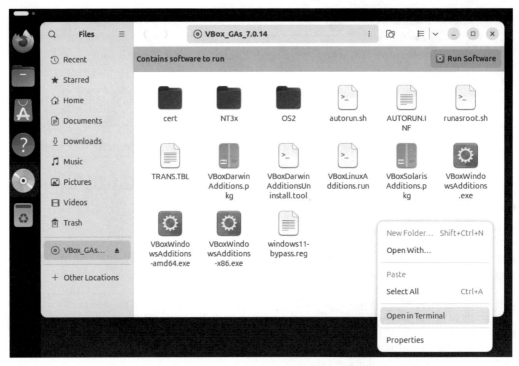

Figure 2.4 – Opening the Guest Additions CD Image folder in the Terminal

You will then be inside a Terminal, where you can find a file named `VBoxLinuxAdditions.run`. You need to run this file with admin privileges:

```
$ sudo ./VBoxLinuxAdditions.run
```

After this, you might get a popup asking you to reboot the VM. Click **Restart Now**. If you don't get this popup, simply run `sudo reboot` in the Terminal.

When the VM starts again, the screen should resize automatically when you change the size of the window. If it doesn't work the first time, I found that running the command once again (`sudo ./VBoxLinuxAdditions.run`) and rebooting may solve the problem.

You can then right-click on the disk and choose **Eject**, and you're all set. To enable copy/paste between the host and the VM, go to the top menu and click **Devices | Shared Clipboard | Bidirectional**.

Now, your VM is fully installed and configured. If you've chosen that path, this will allow you to complete at least *Part 1* and *Part 2* of this book. For *Part 3*, as mentioned previously, you might encounter some issues, especially when using Gazebo. When you reach this point, I suggest that you install Ubuntu natively with a dual boot.

Installing ROS 2 on Ubuntu

Now that you have Ubuntu 24.04 installed on your computer (either as a dual boot or in a VM), let's install ROS 2. As discussed previously, we will install ROS Jazzy here. If you're using a different Ubuntu distribution, make sure to use the appropriate ROS 2 distribution.

The best way to install ROS 2 is to follow the instructions on the official documentation website, for binary packages installation: `https://docs.ros.org/en/jazzy/Installation/Ubuntu-Install-Debians.html`. There are quite a few commands to run, but don't worry—all you need to do is copy and paste them one by one.

> **Note**
> As the installation instructions are often updated, the following commands you see in this book may differ slightly from the ones from the official documentation. If so, copy/paste from the official instructions.

Now, let's start the ROS 2 installation.

Setting the locale

Make sure you have a locale that supports UTF-8:

```
$ locale
```

With this, you can check that you have UTF-8. If in doubt, just run those commands one by one (I do this every time):

```
$ sudo apt update && sudo apt install locales
$ sudo locale-gen en_US en_US.UTF-8
$ sudo update-locale LC_ALL=en_US.UTF-8 LANG=en_US.UTF-8
$ export LANG=en_US.UTF-8
```

Now, check again; you'll see UTF-8 this time:

```
$ locale
```

Setting up the sources

Run those five commands one by one (note that it's better to copy/paste them from the official documentation directly):

```
$ sudo apt install software-properties-common
$ sudo add-apt-repository universe
$ sudo apt update && sudo apt install curl -y
```

```
$ sudo curl -sSL https://raw.githubusercontent.com/ros/rosdistro\
/master/ros.key -o /usr/share/keyrings/ros-archive-keyring.gpg
$ echo "deb [arch=$(dpkg --print-architecture) signed-by=/usr\
/share/keyrings/ros-archive-keyring.gpg] http://packages.ros.org/ros2\
/ubuntu $(. /etc/os-release && echo $UBUNTU_CODENAME) main" | sudo\
tee /etc/apt/sources.list.d/ros2.list > /dev/null
```

The main goal of these commands is to add the ROS packages server to your `apt` sources. Run the following command:

```
$ sudo apt update
```

You should now see additional lines, something like this:

```
Get:5 http://packages.ros.org/ros2/ubuntu noble InRelease [4,667 B]
Get:6 http://packages.ros.org/ros2/ubuntu noble/main amd64 Packages
[922 kB]
```

Those new sources (`packages.ros.org`) will allow you to fetch the ROS 2 packages when installing with `apt`. Here, you can also see `noble`, which means the ROS 2 packages are for Ubuntu Noble (24.04). Since you know how Ubuntu and ROS distributions are linked, this also means ROS Jazzy.

Now that we have the correct sources, we can finally install ROS 2.

Installing ROS 2 packages

As a best practice, upgrade all your existing packages before you install any ROS 2 packages:

```
$ sudo apt update
$ sudo apt upgrade
```

Now, you can install ROS 2 packages. As shown in the documentation, you can choose between **Desktop Install** or **ROS-Base Install**.

ROS 2 is not just one piece of software or package. It's a collection of many packages:

- **ROS-Base**: This contains the bare minimum packages to make ROS 2 work correctly
- **ROS Desktop**: This contains all the packages in ROS-Base, plus lots of additional packages, so you get access to more tools, simulations, demos, and so on

Since you're installing ROS 2 on a computer with a desktop, and you want to have access to as many functionalities as you can, choose ROS Desktop Install. If you were to install ROS on a limited environment, such as Ubuntu Server (no desktop) on a Raspberry Pi board, then ROS-Base would make sense.

Install ROS Desktop by running the following command:

```
$ sudo apt install ros-<distro>-desktop
```

Replace <distro> with the distribution name (using lowercase). So, if you want to install ROS Jazzy, you will use the following command:

```
$ sudo apt install ros-jazzy-desktop
```

As you can see, when you run this command, a few hundred new packages will be installed. This can take some time, depending on your internet connection speed, as well as the performance of your computer (or VM).

If you see an error stating *unable to locate package ros-jazzy-desktop*, for example, it probably means you're trying to install a ROS 2 distribution on the wrong Ubuntu version. This is a common error, so make sure to use the correct Ubuntu/ROS 2 pairing (as seen previously in this chapter).

> **Note**
> From this, it's quite easy to see how to install any other ROS 2 package. You just need to write ros, then the distribution name, and then the package name (using dashes, not underscores): ros-<distro>-package-name. For example, if you want to install the abc_def package on ROS Jazzy, then you will need to run sudo apt install ros-jazzy-abc-def.

Once the installation is done, you can also install the ROS development tools. We'll need these in the remainder of this book to compile our code and create ROS programs. For this command, no ROS distribution needs to be specified; it's the same for all of them:

```
$ sudo apt install ros-dev-tools
```

Once you've done this, ROS 2 will be installed, and you'll have all the ROS 2 tools you need.

I also recommend that you frequently update the ROS 2 packages you've installed. To do so, simply run sudo apt update and sudo apt upgrade, like you would update any other package.

Setting up the environment for ROS 2

At this point, open a Terminal and run the following command:

```
$ ros2
ros2: command not found
```

You will get an error message saying that the ros2 command can't be found. As we will see later, ros2 is a command-line tool we can use to run and test our programs from the Terminal. If this command isn't working, it means that ROS 2 hasn't been set up correctly.

Even if ROS 2 is installed, there's one more thing you need to do in every new session (or Terminal) where you want to use ROS 2: you need to **source** it in the environment.

Sourcing ROS 2 in the environment

To do that, source this bash script from where ROS 2 is installed:

```
$ source /opt/ros/<distro>/setup.bash
```

Replace `<distro>` with the current distribution name you are using. For ROS Jazzy, run the following command:

```
$ source /opt/ros/jazzy/setup.bash
```

After you run this, try executing the `ros2` command again. This time, you should get a different message (usage message). This means that ROS 2 is correctly installed and set up in your environment.

> **Note**
> We will see how to use the `ros2` command line later in this book. For now, you can just use it to see if ROS 2 has been set up correctly or not.

Adding the source line to the .bashrc file

You must source this bash script every time you open a new session or Terminal. To make things easier and so you don't forget about it, let's just add this command line to the `.bashrc` file.

If you don't know what `.bashrc` is, simply put, it's a bash script that will run every time you open a new session (that can be an SSH session, a new Terminal window, and so on). This `.bashrc` file is specific to each user, so you will find it in your home directory (as a hidden file because of the leading dot).

You can add the source line to the `.bashrc` file with this command:

```
$ echo 'source /opt/ros/<distro>/setup.bash' >> ~/.bashrc
```

Replace `<distro>` with your ROS distribution name. You could also just open the `.bashrc` file directly with any text editor—gedit, nano, or Vim—and add the source line at the end. Make sure it's only added once.

Once you've done this, any time you open a new Terminal, you can be sure that this Terminal is correctly sourced, and thus you can use ROS 2 in it.

Now, to make a final check, open a Terminal and run this command (there's no need to understand anything for now; it's just to verify the installation):

```
$ ros2 run turtlesim turtlesim_node
```

This will print a few logs, and you should see a new window with a turtle in the middle. To stop the program, press *Ctrl + C* in the Terminal where you ran the command.

That's it for installing and configuring ROS 2. I will just give you a few more tips on what development tools can be useful when developing with ROS.

Extra tools for ROS development

Apart from the mandatory previous steps to install and set up ROS 2, any other development tool is up to you. If you have your favorite way of working with a Terminal, your favorite text editor, or your favorite **integrated development environment** (**IDE**), this is perfectly fine.

In this section, I will show you a few tools that lots of ROS developers use (me included) and that I think can help you get a better experience when developing with ROS.

Visual Studio Code

Visual Studio Code (**VS Code**) is a quite popular IDE used by many developers. What makes it nice for us is its good support for ROS development.

VS Code is free to use and open source; you can even find its code on GitHub. To install VS Code, open a Terminal and run the following command:

```
$ sudo snap install code --classic
```

The installation just requires one line and uses Ubuntu's Snap feature. After installing it, you can open VS Code by searching for it in the Ubuntu applications, or simply by running `code` in the Terminal.

Now, start VS Code and go to the **Extensions** panel—you can find it on the left menu.

There, you can search for the ROS extension by typing `ros`. There are quite a few; choose the one developed by Microsoft. This extension is compatible with both ROS 1 and ROS 2, so there's no problem here:

Figure 2.5 – The ROS extension to install in VS Code

Install this extension. This will also install a bunch of other extensions, notably Python and C++ extensions, which are quite useful when writing code.

On top of this, I also usually install the **CMake** extension by `twxs` (just type `cmake` and you'll find it). With this, we get nice syntax highlighting when writing into `CMakeLists.txt` files, which is something we will have to do quite often with ROS 2.

The Terminal and other tools

As you develop with ROS 2, you will often need to open several Terminals: one for compiling and installing, a few to run the different programs of your application, and a few more for introspection and debugging.

It can become quite difficult to keep track of all the Terminals you use, so as best practice, it's nice to have a tool that can easily handle multiple Terminals in one window.

There are quite a few tools for doing this. The one I'm going to talk about here is called **terminator**. Not only does it have a funny name, but it's also super practical to use.

To install `terminator`, run the following command:

```
$ sudo apt install terminator
```

Then, you can find it from the applications menu, run it, right-click on the left bar menu, and choose **Pin to Dash** so that it stays there and becomes easy to start.

You can find all the commands for `terminator` online, but here are the most important ones to get started:

- *Ctrl + Shift + O*: split the selected Terminal horizontally.
- *Ctrl + Shift + E*: split the selected Terminal vertically.
- *Ctrl + Shift + X*: make the current Terminal fill the entire window. Use again to revert.
- *Ctrl + Shift + W*: close a Terminal.

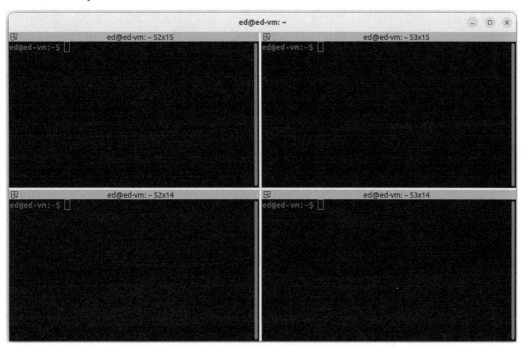

Figure 2.6 – Terminator with four Terminals

Whenever you split a Terminal, this Terminal becomes two different Terminals, each being one session. Thus, you can easily split your Terminal into four or six; this will be enough to run most of your ROS 2 applications. Since we previously added the line to source ROS 2 in the `.bashrc` file, you can use ROS 2 directly in each new Terminal.

Summary

When choosing a ROS 2 distribution, I recommend that you pick the latest LTS distribution, given that it's a few months old and contains all the functionalities you need.

To install and set up ROS 2, you first need to have Ubuntu installed. Each ROS 2 distribution is linked to a specific Ubuntu distribution.

For ROS 2 Jazzy, you must install Ubuntu 24.04. The best option is to have it natively with a dual boot, but to get started quickly, you can also choose to install it in a VM. Then, you can install the ROS 2 packages.

After this, it's important to source the environment for ROS 2 by sourcing a bash script from the ROS installation folder. You can add the line to source this script into your `.bashrc` file so that you don't need to do this every time you open a new Terminal.

Finally, to get a better development experience with ROS, I suggest using VS Code with the ROS extension, and a tool that allows you to split the Terminal into multiple Terminals, such as `terminator`.

With this setup, you are fully ready to start using ROS 2. In the next chapter, you will run your first ROS 2 programs and discover the core concepts.

3
Uncovering ROS 2 Core Concepts

You will now start your first ROS 2 programs. As you will see, a ROS 2 program is called a **node**.

What's inside a node, what does it do, and how do nodes communicate with each other? How do you configure nodes and start several of them at the same time?

That's what we will focus on in this chapter. We won't write any code yet but instead focus on discovering the concepts through hands-on experimentation, using existing demos that were installed along with ROS 2.

By the end of this chapter, you will have a global understanding of the main ROS 2 core concepts. You will also be familiar with the most important ROS 2 tools that you will use later in all your projects.

> **Important note**
>
> In this chapter, I won't explain everything. We are going to embark on a discovery phase, where we use the different core concepts and guess how they work. Not everything has to make sense right now and don't worry too much if some concepts are still a bit blurry for you. Just try to get through the chapter by running all the commands yourself.

The goal here is not to get a complete understanding or to remember all the commands, but rather to get an *intuition* of how things work. This will help you tremendously for *Part 2* when we go through each concept with much more detail—and develop with them.

In this chapter, we will cover the following topics:

- Running your first node
- Topics
- Services

- Actions
- Parameters
- Launch files

Running your first node

To understand what a node is, we will simply run one and make some observations using some of the most useful ROS 2 tools.

For this chapter, I recommend having a few open terminals. You can start a few terminal windows and arrange them on your screen or run Terminator (see *Extra tools for ROS development* in *Chapter 2*) with at least three tabs. To clear any confusion when running a command, I will also tell you in which terminal to run the command (Terminal 1, Terminal 2, etc.).

Starting a node from the terminal with ros2 run

Let's discover your first ROS 2 tool, and probably the most important one: the `ros2` command-line tool. You will use this tool all the time in your future projects.

`ros2` comes with a lot of functions. We will explore some of them in this chapter, and more in the following ones. There is no need to remember all the commands: just use them to build an understanding now, and later you will easily be able to retrieve them from the terminal.

To start a node, you have to follow this template: `ros2 run <package> <executable>`.

As we will see later, nodes are organized inside packages. That's why you first need to specify the package name where the node is and the executable name for that node. As we installed ROS Desktop, a lot of demo packages are already included, for example, `demo_nodes_cpp`.

In Terminal 1, start the talker node from the `demo_nodes_cpp` package:

```
$ ros2 run demo_nodes_cpp talker
[INFO] [1710223859.331462978] [talker]: Publishing: 'Hello World: 1'
[INFO] [1710223860.332262491] [talker]: Publishing: 'Hello World: 2'
[INFO] [1710223861.333233619] [talker]: Publishing: 'Hello World: 3'
^C[INFO] [1710223862.456938986] [rclcpp]: signal_handler(signum=2)
```

After you run this command, the node starts. To stop it, simply press *Ctrl + C* in the terminal where the node is running.

So, what happened here? From what we can observe, this node is simply a program that will print a log in the terminal every second.

Now, keep the node alive, or start it again if you stopped it. In another terminal (Terminal 2), let's start a different node, which is the listener node from the same package:

```
$ ros2 run demo_nodes_cpp listener
[INFO] [1710224252.496221751] [listener]: I heard: [Hello World: 9]
[INFO] [1710224253.497121609] [listener]: I heard: [Hello World: 10]
[INFO] [1710224254.495878769] [listener]: I heard: [Hello World: 11]
```

This node is also a simple program that will print some logs in the terminal. However, as you can see, when the two nodes are running (talker and listener), whatever is printed on the talker seems to also be received on the listener, which then prints it.

In this example, we have two nodes running, and we can clearly see that they communicate with each other. If you stop the talker node, you will see that the listener node stops printing logs as well. When you restart the talker, the listener starts printing what the talker is "sending."

> **Note**
>
> Here are a few tips for when you use the `ros2` command-line tool:
>
> Use auto-completion as much as you can. It will make you type commands faster, but more importantly, you will be sure that you type the right command, package name, node name, and so on.
>
> If you have any doubts about a command or sub-command, you can get help from the terminal by adding `-h` to the command. For example, use `ros2 -h` for the global help, or `ros2 run -h` for help specifically for the run sub-command. There's no need to remember all the commands if you know where to find the information.

Introspecting the nodes with rqt_graph

There is another very useful tool we will discover here, which is a good complement to the command line: `rqt_graph`. This tool will show you all running nodes with a nice visual.

Keep the 2 nodes alive (Terminals 1 and 2) and start `rqt_graph` in Terminal 3. The command is identical to the tool name:

```
$ rqt_graph
```

This will open a new graphical window, where you should see the two nodes. If you don't see anything, make sure both nodes are running, and refresh the view by clicking on the button with a refresh icon, in the top-left corner. You can also select **Nodes/Topics (all)** from the top left drop-down menu. Then, you should get something like this:

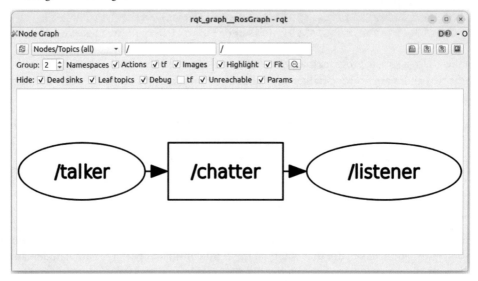

Figure 3.1 – rqt_graph with two nodes

Here, we can see that both nodes are up and running (nothing new for now), but we also see an arrow going from the talker node to a box, and another one from that box to the listener node. This is the ROS 2 communication that allows the talker node to send some data—here, some text—to the listener node. We will talk about this communication in the next section of this chapter.

What we can conclude for now is that we started two different ROS programs (nodes) in two different terminals, using the `ros2 run` command. It seems that those two programs are communicating with each other, and we can confirm that with `rqt_graph`.

Before we go further and look at what kind of ROS communication it is, let's run another set of nodes.

Running a 2D robot simulation

The first two nodes we ran are very simple programs that print logs on the terminal and send some text between each other.

Now, stop all existing nodes (press *Ctrl + C* in each terminal), and let's start again with some other nodes. In Terminal 1, run the following command:

```
$ ros2 run turtlesim turtlesim_node
Warning: Ignoring XDG_SESSION_TYPE=wayland on Gnome. Use QT_QPA_
```

```
PLATFORM=wayland to run on Wayland anyway.
[INFO] [1710229365.273668657] [turtlesim]: Starting turtlesim with
node name /turtlesim
[INFO] [1710229365.288027379] [turtlesim]: Spawning turtle [turtle1]
at x=[5.544445], y=[5.544445], theta=[0.000000]
```

You'll see a few logs, but more importantly, you will get a new window with a blue background and a turtle in the middle. This turtle represents a (very simplified) simulated robot that moves in 2D space.

In Terminal 2, start this second node:

```
$ ros2 run turtlesim turtle_teleop_key
Reading from keyboard
-----------------------------
Use arrow keys to move the turtle.
Use G|B|V|C|D|E|R|T keys to rotate to absolute orientations. 'F' to
cancel a rotation.
'Q' to quit.
```

After you see this, make sure Terminal 2 is selected, and use the arrow keys (up, down, left, right). When doing this, you should see the turtle robot moving.

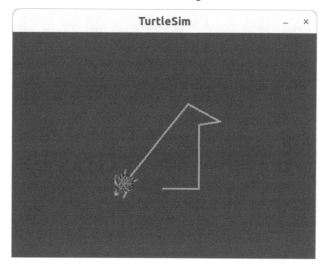

Figure 3.2 – Moving the simulated turtle robot (TurtleSim)

We can already guess that the second node we started (with `turtle_teleop_key`) is reading the keys you press on the keyboard and sending some kind of information/command to the `turtlesim` node, which then makes the turtle robot move. To confirm that, start `rqt_graph` again on Terminal 3:

```
$ rqt_graph
```

If needed, refresh the view a few times. Select **Nodes/Topics (all)**, and you'll see something like this:

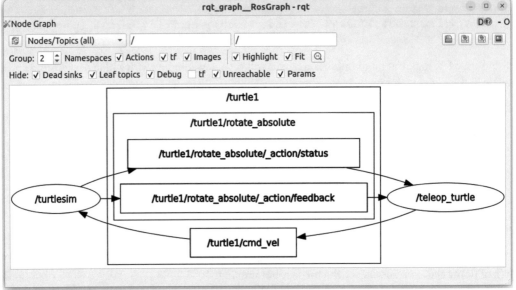

Figure 3.3 – rqt_graph with the turtlesim and teleop_turtle nodes

We can find a node named `turtlesim` and another one named `teleop_turtle`, and we can clearly see that the two nodes are communicating with each other.

> **Note**
>
> As you can see, the executable name (`turtle_teleop_key`) we used to start the node is not necessarily the same as the node name (`teleop_turtle`). We'll come back to this later in this book.

Recap – nodes

What can we get from those two experiments? As you can see, a ROS 2 node can be any kind of computer program that contains:

- Instructions to print logs in the terminal
- Graphical windows (2D, can also be 3D)
- Hardware drivers and more

A node, on top of being a computer program, also benefits from the ROS 2 functionalities: logs, communication with other nodes, and other features we will discover throughout this book.

Now that you've seen how to start a node and use the `ros2` command-line tool, let's focus on how they communicate with each other.

Topics

Nodes communicate with each other using ROS 2 communication features. There are three types of communication: topics, services, and actions. We will discover all three of them, starting with **topics**.

Here, we will make some basic discoveries to get an idea of what a ROS 2 topic is, and you'll learn much more about them, including how to write code for topics, in *Chapter 5*.

Running a topic publisher and subscriber

Stop all running nodes (*Ctrl + C*), and let's come back to our first example.

In Terminal 1, input the following:

```
$ ros2 run demo_nodes_cpp talker
```

In Terminal 2, input the following:

```
$ ros2 run demo_nodes_cpp listener
```

In Terminal 3, input the following:

```
$ rqt_graph
```

If needed, refresh the view a few times, select **Nodes/Topics (all)**, and you should get the same visual as in *Figure 3.1*.

In the middle, you will see a `/chatter` box. This box represents a ROS 2 topic. What you can also see is that the talker node is sending something to the `/chatter` topic, which will then be received by the listener node.

We say that the talker is a **publisher**, and the listener is a **subscriber**.

An important detail is that the talker is not actually sending data directly to the listener. The talker is publishing on the `/chatter` topic, and the listener is subscribing to the `/chatter` topic. Because of this, the data flows from the talker to the listener.

A name and an interface (data type)

From `rqt_graph`, we can already see that one node can send data to another node through a topic. An important point is that the topic is defined by a name. Both publishers and subscribers use the same name to make communication successful.

There is more to it than just a name. Let's come back to the terminal and use the `ros2` command-line tool to discover more information.

You previously used `ros2 run` to start a node. We also have `ros2 topic` to interact with topics. You can get more help with all available topic commands with `ros2 topic -h`. The `ros2 topic list` command will list all available topics, which means all topic communications between running nodes.

In Terminal 3 (if you stopped `rqt_graph`), or in Terminal 4, run the following:

```
$ ros2 topic list
/chatter
/parameter_events
/rosout
```

For any node you create, you will always see `/rosout` and `/parameter_events`. Those are not important for now and you can just ignore them. What's important is the `/chatter` topic. We already know it's used between the talker and listener node, but now the question is this: What kind of data is being sent?

To get this information, we can use `ros2 topic info <topic_name>`:

```
$ ros2 topic info /chatter
Type: std_msgs/msg/String
Publisher count: 1
Subscription count: 1
```

Here, we see how many nodes are publishing and subscribing to this topic. We have one publisher (talker node) and one subscriber (listener node). We can also see what kind of message is being sent: `std_msgs/msg/String`. In ROS 2, this message is called an **interface**.

To see what's inside an interface, run `ros2 interface show <interface_name>`:

```
$ ros2 interface show std_msgs/msg/String
# Some comments
string data
```

There can be a bunch of comments (starting with #) that you can ignore. The important thing is this: `string data`. This tells us what's being sent on that topic. Here, it's a string (chain of characters) with the name `data`.

So, when the talker node wants to send a message to the `/chatter` topic, it needs to send a `data` field of type string. The listener, to get that information, will need to subscribe to `/chatter`, and expect to receive the same data type.

That is how a topic is defined: a name and an interface (data type). Both publishers and subscribers should use the same name and interface to communicate.

This makes sense: as an analogy, imagine you and I are talking through an online chat. If we are not in the same chat room (same topic name), we won't be able to find each other. Also, if I'm talking to you in a language you don't speak, this would not make sense to you. To communicate, we both need to agree on what language we use (same interface).

More experimentation with topics

Let's practice a bit more with a challenge. This time, I won't just show you what commands to run directly but give you a challenge so you can practice on your own.

> **Note**
> I will sometimes give you some challenges/activities in this book with various levels of difficulty —this first challenge being quite small. I will then give you the solution (or a part of it). Of course, I encourage you to stop reading after the instructions and only use the previous pages to solve the challenge. Then, read the solution and compare it with what you did.

Challenge

Run the second example we did with the 2D `turtlesim` robot (two nodes).

What I challenge you to do right now is to find the topic name and interface used by the `turtle_teleop` node to send a velocity command to the `turtlesim` node. Use the previous commands from this chapter to try to get that information.

Solution

Start the two nodes and `rqt_graph`.

In Terminal 1, input the following:

```
$ ros2 run turtlesim turtlesim_node
```

In Terminal 2, input the following:

```
$ ros2 run turtlesim turtle_teleop_key
```

In Terminal 3, input the following:

```
$ rqt_graph
```

Make sure you refresh the view on `rqt_graph` and select **Nodes/Topics (all)**. You will get the same as previously, as in *Figure 3.3*.

There are some more things on this screen, but we just need one piece of information. As you can see, there is a /turtle1/cmd_vel box, here representing a topic. The teleop_turtle node is a publisher, and the turtlesim node is a subscriber to that topic.

This is quite logical: the teleop_turtle node will read the keys that you press, and then publish on the topic. On its end, the turtlesim node will subscribe to that topic to get the latest velocity command for the robot.

We can get roughly the same information from the terminal:

```
$ ros2 topic list
/parameter_events
/rosout
/turtle1/cmd_vel
/turtle1/color_sensor
/turtle1/pose
```

From the list of all running topics, we can spot the /turtle1/cmd_vel topic.

Now, to retrieve the interface (data type) for that topic, run the following command:

```
$ ros2 topic info /turtle1/cmd_vel
Type: geometry_msgs/msg/Twist
Publisher count: 1
Subscription count: 1
```

For the details of what's inside the interface, run the following command:

```
$ ros2 interface show geometry_msgs/msg/Twist
# This expresses velocity in free space broken into its linear and
angular parts.
Vector3  linear
        float64 x
        float64 y
        float64 z
Vector3  angular
        float64 x
        float64 y
        float64 z
```

This interface is a bit more complex than the one we had before. There's no need to make sense of all of that for now, as we will dive into interfaces later in this book. The goal here is just to find a topic name and interface.

From this information, let's say we want to make the robot move forward. We can guess that we would need to set a value for the x field, inside the linear field (as x is pointing forward in ROS).

Recap – topics

With those two experiments, you can see that nodes communicate with each other using topics. One node can publish or subscribe to a topic. When publishing, the node sends some data. When subscribing, it receives the data.

A topic is defined by a name and a data type. That's all you need to remember for now. Let's switch to the second communication type: services.

Services

Topics are very useful to send a stream of data/commands from one node to another node. However, this is not the only way to communicate. You can also find client/server communications in ROS 2. In this case, **services** will be used.

As we did for topics, we will run two nodes communicating with each other, this time with services, and we will try to analyze, using the ROS 2 tools, what's happening and how the communication is working.

In *Chapter 6*, you will get a much more detailed explanation about services, when to use them versus topics, and how to include them in your code. For now, let's just continue with our discovery phase.

Running a service server and client

Stop all running nodes. This time, we will start another node from `demo_nodes_cpp`, which contains a simple service server to add two integer numbers. We will also start a **client** node, which will send a request to the **server** node.

In Terminal 1, input the following:

```
$ ros2 run demo_nodes_cpp add_two_ints_server
```

In Terminal 2, input the following:

```
$ ros2 run demo_nodes_cpp add_two_ints_client
```

As soon as you run the client node, you can see this log in Terminal 1 (server):

```
[INFO] [1710301834.789596504] [add_two_ints_server]: Incoming request
a: 2 b: 3
```

You can also see this log in Terminal 2 (client):

```
[INFO] [1710301834.790073100] [add_two_ints_client]: Result of add_
two_ints: 5
```

From what we observe here, it seems that the server node is hanging and waiting. The client node will send a request to the server, with two integer numbers, in this example: 2 and 3. The server node receives the request, adds the number, and returns the result: 5. Then, the client gets the response and prints the result.

This is basically how a service works in ROS 2. You run one node that contains a server, then any other node (client) can send a request to that server. The server processes the request and returns a response to the client node.

A name and an interface (data type)

As for topics, services are defined by two things: a name, and an interface (data type). The only difference is that the interface will contain two parts: a **request** and a **response**.

Unfortunately, `rqt_graph` does not support service introspection—although there are some plans to implement this in future ROS 2 distributions.

To find the name of the service, we can use the `ros2` command-line tool again, this time with the `service` command, followed by `list`. As you can see, if you understand the way to list all topics, then it's exactly the same for services.

At this point, you still have the service node running on Terminal 1, and nothing running on Terminal 2 (as the client stopped after receiving the response). In Terminal 2 or 3, run the following:

```
$ ros2 service list
/add_two_ints
/add_two_ints_server/describe_parameters
/add_two_ints_server/get_parameter_types
/add_two_ints_server/get_parameters
/add_two_ints_server/list_parameters
/add_two_ints_server/set_parameters
/add_two_ints_server/set_parameters_atomically
```

That's a lot of services. Most of them can be discarded. For each node, you automatically get six additional services, all of them containing the name `parameter`. If we ignore them, we can see the /add_two_ints service, which is the service server running on the add_two_ints_server node.

Great, we found the name. Now, to get the data type, we can use `ros2 service type <service_name>`, and then `ros2 interface show <interface_name>`:

```
$ ros2 service type /add_two_ints
example_interfaces/srv/AddTwoInts

$ ros2 interface show example_interfaces/srv/AddTwoInts
int64 a
int64 b
```

```
---
int64 sum
```

You can see that the interface contains a line with three dashes (- - -). This is the separation between the request and the response. With this, you know that to send a request to the server (as a client), you need to send one integer number named a, and another integer number named b. Then, you will receive a response containing one integer number named sum.

Sending a request from the terminal

Instead of running the `add_two_ints_client` node, we can also send a request directly from the terminal. I'm adding this here because it's a very useful way to test a service without requiring an existing client node.

The syntax is `ros2 service call <service_name> <interface_name> "<request_in_json>"`. As you can see, we need to provide both the service name and interface.

Here is an example of how to do that (make sure the server node is still running):

```
$ ros2 service call /add_two_ints example_interfaces/srv/AddTwoInts
"{a: 4, b: 7}"
waiting for service to become available...
requester: making request: example_interfaces.srv.AddTwoInts_
Request(a=4, b=7)
response:
example_interfaces.srv.AddTwoInts_Response(sum=11)
```

With this command, we send a request with 4 and 7. The server node will print those logs:

```
[INFO] [1710302858.634838573] [add_two_ints_server]: Incoming request
a: 4 b: 7
```

In the end, on the client side, we get the response that contains `sum=11`.

More experimentation with services

Here's another challenge for you to practice with services.

Challenge

Start the `turtlesim` node, list the existing services, and find how to spawn a new turtle robot in the 2D screen, using the terminal.

Once again, I recommend you take a bit of time to try to do this on your own. Feel free to review all the previous commands from this chapter. No need to remember all of them as you can easily find them in the book, using the *Tab* key for auto-completion, or by adding `-h` to any command.

Solution

Stop all running nodes.

In Terminal 1, input the following:

```
$ ros2 run turtlesim turtlesim_node
```

In Terminal 2, input the following:

```
$ ros2 service list
/clear
/kill
/reset
/spawn
/turtle1/set_pen
/turtle1/teleport_absolute
/turtle1/teleport_relative
# There are more services containing "parameter" that we can ignore
```

Those are all the services we can use for the `turtlesim` node. As you can see, we already have quite a lot. In this challenge, you have to spawn a turtle. Great, we can find a /spawn service.

We already have the name; now, let's find the interface (request, response):

```
$ ros2 service type /spawn
turtlesim/srv/Spawn
$ ros2 interface show turtlesim/srv/Spawn
float32 x
float32 y
float32 theta
string name # Optional.  A unique name will be created and returned if this is empty
---
string name
```

Now, we have all the information we need. To send a request to the server, we have to use the /spawn service and the `turtlesim/srv/Spawn` interface. We can send a request that contains (x, y, theta) coordinates, plus an optional name. Actually, note that all fields in the request are optional. If you don't provide a value for a field, the default will be 0 for numbers, and " " for strings.

Let's now send our request from the terminal:

```
$ ros2 service call /spawn turtlesim/srv/Spawn "{x: 3.0, y: 4.0}"
waiting for service to become available...
requester: making request: turtlesim.srv.Spawn_Request(x=3.0, y=4.0, theta=0.0, name='')
response:
turtlesim.srv.Spawn_Response(name='turtle2')
```

If you look at the 2D window, you will see a new turtle.

Figure 3.4 – The TurtleSim window after spawning a new turtle

This turtle has been spawned at the (x, y, theta) coordinates provided in the request. You can try to run the `ros2 service call` command again a few times with different coordinates, so you can spawn more turtles on the screen.

Recap – services

You have successfully run a client/server communication between two nodes. Once again, a service is defined by a name and an interface (request, response).

For more details about the question of when to use topics versus services, read on, as this is something we will see later in this book when you understand more about each concept. For now, you have just seen two kinds of communication between nodes. Each of them has a name and an interface, and we can already play with them in the terminal.

There is now one more ROS 2 communication to discover: actions.

Actions

A ROS 2 **action** is basically the same thing as a service (client/server communication), but designed for longer tasks, and when you might want to also get some feedback during the execution, be able to cancel the execution, and so on.

In robotics, we are making robots move. Making a robot move is not something that happens instantly. It could take a fraction of a second, but sometimes a task could take a few seconds/minutes or more. ROS 2 services have been designed for quick execution, for example: a computation, or an immediate action, such as spawning a turtle on a screen. Actions are used whenever a client/server communication might take more time and we want more control over it.

We will dive into actions with more details in *Chapter 7*. Actions are what I consider to be an intermediate-level concept, not a beginner one, so I won't start to go too deep right now. Let's just continue the discovery phase with a very simple example, just to get an idea of how it works.

Running an action server

Stop all running nodes, and start the `turtlesim` node again in Terminal 1:

```
$ ros2 run turtlesim turtlesim_node
```

As you've already practiced with topics and services, the following `ros2` commands will start to look familiar to you. List all existing actions in Terminal 2:

```
$ ros2 action list
/turtle1/rotate_absolute
```

From what we observe, it seems that the `turtlesim` node contains an action server named `/turtle1/rotate_absolute`. There is no existing client node for this action, so we will try to interact with it from the terminal. Of course, we will need two things: the name and the interface.

A name and an interface (data type)

As for topics and services, an action will be defined by a name and an interface. This time, the interface contains three parts: **goal**, **result**, and **feedback**.

The goal and result are similar to the request and response for a service. The feedback is additional data that can be sent by the server to give some feedback during the goal execution.

To get the action interface, you can run the `ros2 action info <action_name> -t` command. Don't forget to add `-t` (for type), otherwise, you'll see some details, but no interface:

```
$ ros2 action info /turtle1/rotate_absolute -t
Action: /turtle1/rotate_absolute
Action clients: 0
Action servers: 1
    /turtlesim [turtlesim/action/RotateAbsolute]
```

We can see that the action is running within one server (the `turtlesim` node), and we also found the interface: `turtlesim/action/RotateAbsolute`.

Let's see what's inside this interface:

```
$ ros2 interface show turtlesim/action/RotateAbsolute
# The desired heading in radians
float32 theta
---
# The angular displacement in radians to the starting position
float32 delta
---
# The remaining rotation in radians
float32 remaining
```

You can see two separations with three dashes (- - -). The first part is the goal, the second part is the result, and the third part is the feedback. This action is quite simple; we only have one float number for each part of the interface.

As a client, we send the desired angle for rotation. The server node will receive the goal and process it while optionally sending some feedback. When the goal is finished, the server will send the result to the client.

Sending a goal from the terminal

As an action client, we are firstly interested in the goal part of the interface. Here, we need to send a float number, which corresponds to the angle (in radians) we want to rotate the turtle to.

The syntax to send a goal from the terminal is `ros2 action send_goal <action_name> <action_interface> "<goal_in_json>"`. Once again, you need to provide both the name and interface.

Make sure the `turtlesim` node is alive, then send a goal from Terminal 2:

```
$ ros2 action send_goal /turtle1/rotate_absolute turtlesim/action/RotateAbsolute "{theta: 1.0}"
Waiting for an action server to become available...
Sending goal:
     theta: 1.0
Goal accepted with ID: 3ba92096282a4053b552a161292afc8e
Result:
    delta: -0.9919999837875366
Goal finished with status: SUCCEEDED
```

After you run the command, you should see the turtle robot rotate on the 2D window. Once the desired angle is reached, the action will finish, and you will receive the result.

Recap – actions

You have run your first action communication in ROS 2. An **action** is defined by two things: a name and an interface (goal, result, feedback). Actions are used when you need a client/server kind of communication, and when the duration of the action might take some time—versus being executed immediately.

With this, you have seen all three types of communications in ROS 2: topics, services, and actions. Each one will get its own chapter in *Part 2* so you can see in detail how they work, how to use them in your code, and how to fully introspect them with ROS 2 tools.

Parameters

We are now going to come back to the node itself and talk about another important ROS 2 concept: **parameters**.

This time, it's not about communication, but about how to give different settings to a node when you start it.

Let's quickly discover how parameters work, and you'll get a complete explanation with more examples and use cases in *Chapter 8*.

Getting the parameters for a node

Stop all running nodes, and start the `turtlesim` node in Terminal 1:

```
$ ros2 run turtlesim turtlesim_node
```

Then, to list all parameters, it's quite easy, and you can probably guess the command. If we have `ros2 topic list` for topics, `ros2 service list` for services, and `ros2 action list` for actions, then, for parameters, we have `ros2 param list`. The only particularity is that we use the word `param` instead of `parameter`. Run this command in Terminal 2:

```
$ ros2 param list
/turtlesim:
  background_b
  background_g
  background_r
  holonomic
  qos_overrides./parameter_events.publisher.depth
  qos_overrides./parameter_events.publisher.durability
  qos_overrides./parameter_events.publisher.history
  qos_overrides./parameter_events.publisher.reliability
  start_type_description_service
  use_sim_time
```

> **Note**
>
> Sometimes, the `ros2 param list` command doesn't work properly and you won't see any parameters or not all of them. This can also happen with a few other `ros2` commands. In this case, just run the command again, a few times if needed, and this should work. It's probably some kind of bug in the `ros2` command-line tool itself, but nothing to worry about: the application is running correctly.

We first see the `turtlesim` node (actually written `/turtlesim`, with a leading slash), then a list of names under this node, with an indentation. Those names are the parameters, and they belong to the node. That's the first thing about parameters in ROS 2: they exist within a node. If you stop this `turtlesim` node, then the parameters would also be destroyed.

There are a bunch of parameters you can ignore: `use_sim_time`, `start_type_description_service`, and all the parameters containing `qos_overrides`. Those will be present for any node you start. If we get rid of them, we are left with a few parameters, including `background_b`, `background_g`, `backgound_r`.

From this observation, it seems that we would be able to change the background color of the 2D window when we start the `turtlesim` node.

Now, what's inside those parameters? What kind of value? Is it a round number, a float, or a string? Let's find out, with `ros2 param get <node_name> <param_name>`. In Terminal 2, run the following commands:

```
$ ros2 param get /turtlesim background_b
Integer value is: 255
$ ros2 param get /turtlesim background_g
Integer value is: 86
$ ros2 param get /turtlesim background_r
Integer value is: 69
```

From this, we can guess that the **red, green, blue** (**RGB**) value for the background is (69, 86, 255). It also seems that the parameter value is a round number from 0 to 255.

Setting up a parameter value for a node

Now that we have found the name of each parameter, and what kind of value we should use, let's modify the value ourselves when we start the node.

For this, we will need to restart the node, using the same syntax as before: `ros2 run <package_name> <executable_name>`. We will then add `--ros-args` (only once), and `-p <param_name>:=value` for each parameter we want to modify.

Stop the `turtlesim` node on Terminal 1, and start it again, with a different value for some of the parameters:

```
$ ros2 run turtlesim turtlesim_node --ros-args -p background_b:=0 -p
background_r:=0
```

Here, we decided that both the blue and red colors would be 0. We don't specify any value for `background_g`, which means that the default value will be used (as seen previously: 86).

After you run this command, you should see the 2D screen appear, but this time, the background is dark green.

Recap – parameters

Parameters are settings that can be provided at runtime (which means when we run the node). They allow us to easily configure the different nodes that we start, and thus, they make ROS 2 applications more dynamic.

A parameter exists within a node. You can find all parameters for a node and get the value for each one. When starting the node, you can give a custom value for the parameters you want to modify.

Launch files

Let's finish this list of ROS 2 concepts with launch files.

A **launch file** will allow you to start several nodes and parameters from just one file, which means that you can start your entire application with just one command line.

In *Chapter 9*, you will learn how to write your own launch file, but for now, let's just start a few to see what they do.

Starting a launch file

To start a single node in the terminal, you have seen the `ros2 run` command. For launch files, we will use `ros2 launch <package_name> <launch_file>`.

Stop all running nodes, and let's start the `talker_listener` launch file from the `demo_nodes_cpp` package. In Terminal 1, run the following command:

```
$ ros2 launch demo_nodes_cpp talker_listener_launch.py
[INFO] [launch]: All log files can be found below /home/ed/.ros/
log/2024-03-14-16-09-27-384050-ed-vm-2867
[INFO] [launch]: Default logging verbosity is set to INFO
[INFO] [talker-1]: process started with pid [2868]
[INFO] [listener-2]: process started with pid [2871]
```

```
[talker-1] [INFO] [1710403768.481156318] [talker]: Publishing: 'Hello
World: 1'
[listener-2] [INFO] [1710403768.482142732] [listener]: I heard: [Hello
World: 1]
```

As you can see, it seems that both the talker and listener nodes have been started. You can easily verify that in Terminal 2:

```
$ ros2 node list
/listener
/talker
```

With `rqt_graph`, you could also check that the nodes communicate with each other. We have proof of that with the logs: on the same screen, we get both logs from the talker and listener nodes, and it seems that the listener node is receiving messages (using the /chatter topic as we saw previously).

In the end, it's the same thing as if we had started both nodes on two terminals. The launch file will simply start the two nodes in one terminal.

If we read the logs more carefully, we can see that each node will be started in a different process. To stop the launch file, press *Ctrl + C*. This will stop all processes (nodes), and your application will end.

Let's now try another launch file from the `turtlesim` package. Stop the launch file in Terminal 1, and start the `multisim` launch file from the `turtlesim` package:

```
$ ros2 launch turtlesim multisim.launch.py
[INFO] [launch]: All log files can be found below /home/ed/.ros/
log/2024-03-14-16-14-41-043618-ed-vm-2962
[INFO] [launch]: Default logging verbosity is set to INFO
[INFO] [turtlesim_node-1]: process started with pid [2963]
[INFO] [turtlesim_node-2]: process started with pid [2965]
```

With this, you will see not one, but two 2D windows, containing each a turtle robot. As you can see from the logs, we are starting two `turtlesim` nodes (two identical nodes with a different name each).

We can also check that from the terminal:

```
$ ros2 node list
/turtlesim1/turtlesim
/turtlesim2/turtlesim
```

The nodes have been renamed. Instead of just /turtlesim, we get /turtlesim1/turtlesim and /turtlesim2/turtlesim. Those names have been chosen inside the launch file.

Recap – launch files

Launch files are quite useful for starting several nodes (and the parameters for those nodes) from one file. With just one command line (`ros2 launch`), you can start an entire ROS 2 application.

There is not much more to say about launch files for now, as this concept is quite simple (the real challenge is when writing a launch file, not starting it). We have now finished discovering the main ROS 2 concepts.

Summary

With this chapter, you have discovered the most important ROS 2 concepts: nodes, topics, services, actions, parameters, and launch files.

ROS 2 programs are called nodes. Simply put, they are regular software programs that can also benefit from ROS 2 functionalities: logs, communications, parameters, and so on.

There are three types of communication: topics, services, and actions. Topics are used to send a stream of data/commands from one or several nodes to another or several other nodes. Services are used when we need client/server communication. Actions are basically the same things as services, but for goal executions that could take some time.

On top of communication features, nodes can also use parameters to specify settings at runtime. Parameters allow nodes to be easily configured when started.

Finally, we can start all nodes and parameters from just one command line, using a launch file.

That's it for the core concepts (for now). You have also discovered the `ros2` command-line tool and `rqt_graph`. Those tools are invaluable, and you will use them all the time. The experiments we did with those tools here are very similar to what you will do in the future for your own ROS 2 projects.

This chapter was a bit special, in a way that it doesn't fully explain one concept from A to Z. As stated in the introduction, it was more of a concept walkthrough, where you discover the main concepts through hands-on discovery. What you get is not a complete understanding, but an intuition of how things work, a bit of experience with the tools, and an idea of the big picture.

Feel free to come back to this chapter and run the experiments again as you make progress with the book. Everything will make much more sense.

You are now ready to continue with *Part 2*, where you will create a complete ROS 2 application from scratch, using Python and C++ code. Each concept you've seen so far will get its own dedicated chapter. The intuition you've developed here will be extremely useful.

Part 2: Developing with ROS 2 – Python and C++

This second part focuses on writing code with ROS 2 and building scalable robotics applications. You already discovered the main concepts in *Chapter 3* and got an intuition of how they work. You will now dive into each concept, one by one, with a real-life analogy, a deep dive into the code (with Python and C++), and extra challenges to help you practice.

This part contains the following chapters:

- *Chapter 4, Writing and Building a ROS 2 Node*
- *Chapter 5, Topics – Sending and Receiving Messages between Nodes*
- *Chapter 6, Services – Client/Server Interaction between Nodes*
- *Chapter 7, Actions – When Services Are Not Enough*
- *Chapter 8, Parameters – Making Nodes More Dynamic*
- *Chapter 9, Launch Files – Starting All Your Nodes at Once*

4
Writing and Building a ROS 2 Node

To write your own custom code with ROS 2, you will have to create ROS 2 programs, or in other words, nodes. You already discovered the concept of nodes in *Chapter 3*. In this chapter, we will go deeper, and you will write your first node with Python and C++.

Before you create a node, there is a bit of setup to do: you need to create a ROS 2 workspace, in which you will build your application. In this workspace, you will then add packages to better organize your nodes. Then, in those packages, you can start to write your nodes. After you write a node, you will build it and run it.

We will do this complete process together, with hands-on code and command lines all along the way. This is the process that you will repeat for any new node you create when developing a ROS 2 application.

By the end of this chapter, you will be able to create your own packages and ROS 2 nodes with Python and C++. You will also be able to run and introspect your nodes from the terminal. This is the stepping stone you need in order to learn any other ROS 2 functionality. There is no topic, service, action, parameter, or launch file without nodes.

All explanations will start with Python, followed by C++, which we'll cover more quickly. If you only want to learn with Python, you can skip the C++ sections. However, if you want to learn with C++, reading the previous Python explanations is mandatory for comprehension.

All the code examples for this chapter can be found in the `ch4` folder of the book's GitHub repository (`https://github.com/PacktPublishing/ROS-2-from-Scratch`).

In this chapter, we will cover the following topics:

- Creating and setting up a ROS 2 workspace
- Creating a package

- Creating a Python node
- Creating a C++ node
- Node template for Python and C++ nodes
- Introspecting your nodes

Technical requirements

To follow this chapter, you need the following:

- Ubuntu 24.04 installed (dual boot or virtual machine)
- ROS Jazzy
- A text editor or IDE (for example, VS Code with the ROS extension)

These requirements will be valid for all chapters in *Part 2*.

Creating and setting up a ROS 2 workspace

Before we write any code, we need to do a bit of organization. Nodes will exist within packages, and all your packages will exist within a **ROS 2 workspace**.

What is a ROS 2 workspace? A **workspace** is nothing more than a folder organization in which you will create and build your packages. Your entire ROS 2 application will live within this workspace.

To create one, you have to follow certain rules. Let's create your first workspace step by step and correctly set it up.

Creating a workspace

To create a workspace, you will simply create a new directory inside your home directory.

As for the workspace's name, let's keep it simple for now and use something that is recognizable: `ros2_ws`.

> **Note**
> The name of the workspace is not important, and it will not affect anything in your application. As we are just getting started, we only have one workspace. When you make progress and start to work on several applications, the best practice is to name each workspace with the name of the application or robot. For example, if you create a workspace for a robot named ABC V3, then you can name it `abc_v3_ws`.

Open a terminal, navigate to your home directory, and create the workspace:

```
$ cd
$ mkdir ros2_ws
```

Then, enter the workspace and create a new directory named `src`. This is where you will write all the code for your ROS 2 application:

```
$ cd ros2_ws/
$ mkdir src
```

That's really all there is to it. To set up a new workspace, you just create a new directory (somewhere in your home directory) and create an `src` directory inside it.

Building the workspace

Even if the workspace is empty (we have not created any packages yet), we can still build it. To do that, follow these steps:

1. Navigate to the workspace root directory. Make sure you are in the right place.
2. Run the `colcon build` command. `colcon` is the build system in ROS 2, and it was installed when you installed the `ros-dev-tools` packages in *Chapter 2*.

Let's build the workspace:

```
$ cd ~/ros2_ws/
$ colcon build
Summary: 0 packages finished [0.73s]
```

As you can see, no packages were built, but let's list all directories under `~/ros2_ws`:

```
$ ls
build   install   log   src
```

As you can see, we have three new directories: `build`, `install`, and `log`. The `build` directory will contain the intermediate files required for the overall build. In `log`, you will find logs for each build. The most important directory for you is `install`, which is where all your nodes will be installed after you build the workspace.

> **Note**
>
> You should always run `colcon build` from the root of your workspace directory, not from anywhere else. If you make a mistake and run this command from another directory (let's say, from the `src` directory of the workspace, or inside a package), simply remove the new `install`, `build`, and `log` directories that were created in the wrong place. Then go back to the workspace root directory and build again.

Sourcing the workspace

If you navigate inside the newly created `install` directory, you can see a `setup.bash` file:

```
$ cd install/
$ ls
COLCON_IGNORE        _local_setup_util_ps1.py    setup.ps1
local_setup.bash     _local_setup_util_sh.py     setup.sh
local_setup.ps1      local_setup.zsh             setup.zsh
local_setup.sh       setup.bash
```

This might look familiar. If you remember, after we installed ROS 2, we sourced a similar bash script from the ROS 2 installation directory (`/opt/ros/jazzy/setup.bash`) so that we could use ROS 2 in our environment. We will need to do the same for our workspace.

Every time you build your workspace, you have to source it so that the environment (the session you are in) knows about the new changes in the workspace.

To source the workspace, source this `setup.bash` script:

```
$ source ~/ros2_ws/install/setup.bash
```

Then, as we previously did, we are going to add that line into our `.bashrc`. This way, you don't need to source the workspace every time you open a new terminal.

Open your `.bashrc` (located in your home directory the path is `~/.bashrc`) using any text editor you want:

```
$ gedit ~/.bashrc
```

Add the line to source the workspace's `setup.bash` script, just after the one to source the global ROS 2 installation. The order is very important here. You have to source the global ROS 2 installation first, and then your workspace, not the other way around:

```
source /opt/ros/jazzy/setup.bash
source ~/ros2_ws/install/setup.bash
```

Make sure to save `.bashrc`. Now, both ROS 2 and your workspace will be sourced in any new terminal you open.

> **Note**
> If you build the workspace in an already sourced environment, you will still need to source the workspace once again as there have been some changes, and the environment is not aware of that. In this case, you can either source the workspace's `setup.bash` script directly, source the `.bashrc`, or open a new terminal.

Your workspace is now correctly set up, and you can build your application. Next step: creating a package.

Creating a package

Any node you create will exist within a package. Hence, to create a node, you first have to create a package (inside your workspace). You will now learn how to create your own packages, and we will see the differences between Python and C++ packages.

But first, what exactly is a package?

What is a ROS 2 package?

A ROS 2 package is a sub-part of your application.

Let's consider a robotic arm that we want to use to pick up and place objects. Before we create any node, we can try to split this application into several sub-parts, or packages.

We could have one package to handle a camera, another package for the hardware control (motors), and yet another package to compute motion planning for the robot.

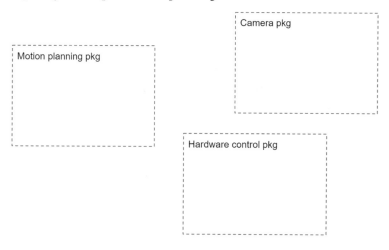

Figure 4.1 – Example of a package organization for a robot

Each package is an independent unit, responsible for one sub-part of your application.

Packages are very useful for organizing your nodes, and also to correctly handle dependencies, as we will see later in this book.

Now, let's create a package, and here you have to make a choice. If you want to create a node with Python, you will create a Python package, and if you want to create a node with C++, you will create a C++ package. The architecture for each package type is quite different.

Creating a Python package

You will create all your packages in the `src` directory of your ROS 2 workspace. So, make sure to navigate to this directory before you do anything else:

```
$ cd ~/ros2_ws/src/
```

Here is how to construct the command to create a package:

1. `ros2 pkg create <pkg_name>`: This is the minimum you need to write.
2. You can specify a build type with `--build_type <build_type>`. For a Python package, we need to use `ament_python`.
3. You can also specify some optional dependencies with `--dependencies <list_of_dependencies_separated_with_spaces>`. It's always possible to add dependencies later in the package.

Let's create our first package named `my_py_pkg`. We will use this name as an example to work with the main ROS 2 concepts. Then, as we progress, we will use more meaningful names. In the `src` directory of your workspace, run the following:

```
$ ros2 pkg create my_py_pkg --build-type ament_python --dependencies rclpy
```

With this command, we say that we want to create a package named `my_py_pkg`, with the `ament_python` build type, and we specify one dependency: `rclpy`—this is the Python library for ROS 2 that you will use in every Python node.

This will print quite a few logs, showing you what files have been created. You might also get a `[WARNING]` log about a missing license, but as we have no intention of publishing this package anywhere, we don't need a license file now. You can ignore this warning.

You can then see that there is a new directory named `my_py_pkg`. Here is the architecture of your newly created Python package:

```
/home/<user>/ros2_ws/src/my_py_pkg
├── my_py_pkg
│   └── __init__.py
├── package.xml
├── resource
│   └── my_py_pkg
├── setup.cfg
├── setup.py
└── test
    ├── test_copyright.py
    ├── test_flake8.py
    └── test_pep257.py
```

Not all the files are important right now. We'll see how to use those files to configure and install our nodes just a bit later.

Here is a quick overview of the most important files and directories:

- `my_py_pkg`: As you can see, inside the package, there is another directory with the same name. This directory already contains an `__init__.py` file. This is where we will create our Python nodes.
- `package.xml`: Every ROS 2 package (Python or C++) must contain this file. We will use it to provide more information about the package as well as dependencies.
- `setup.py`: This is where you will write the instructions to build and install your Python nodes.

Creating a C++ package

We will work a lot with Python in this book, but for completeness, I will also include C++ code for all examples. They will either be explained in the book, or the code will be in the GitHub repository.

Creating a C++ package is very similar to creating a Python package; however, the architecture of the package will be quite different.

Make sure you navigate to the `src` directory of your workspace, and then create a new package. Let's use a similar pattern as we did for Python and name the package `my_cpp_pkg`:

```
$ cd ~/ros2_ws/src/
$ ros2 pkg create my_cpp_pkg --build-type ament_cmake --dependencies rclcpp
```

We choose `ament_cmake` for the build type (meaning this will be a C++ package), and we specify one dependency: `rclcpp`—this is the C++ library for ROS 2, which we will use in every C++ node.

Once again, you should see quite a few logs, with the newly created files, and maybe a warning about the license that you can ignore.

The architecture of your new C++ package will look like this:

```
/home/ed/ros2_ws/src/my_cpp_pkg/
├── CMakeLists.txt
├── include
│   └── my_cpp_pkg
├── package.xml
└── src
```

Here is a quick explanation of the role of each file or directory:

- `CMakeLists.txt`: This will be used to provide instructions on how to compile your C++ nodes, create libraries, and so on.
- `include` directory: In a C++ project, you may split your code into implementation files (`.cpp` extension) and header files (`.hpp` extension). If you split your C++ nodes into `.cpp` and `.hpp` files, you will put the header files inside the `include` directory.
- `package.xml`: This file is required for any kind of ROS 2 package. It contains more information about the package, and dependencies on other packages.
- `src` directory: This is where you will write your C++ nodes (`.cpp` files).

Building a package

Now that you've created one or more packages, you can build them, even if you don't have any nodes in the packages yet.

To build the packages, go back to the root of your ROS 2 workspace and run `colcon build`. Once again, and as seen previously in this chapter, where you run this command is very important.

```
$ cd ~/ros2_ws/
$ colcon build
Starting >>> my_cpp_pkg
Starting >>> my_py_pkg
Finished <<< my_py_pkg [1.60s]
Finished <<< my_cpp_pkg [3.46s]
Summary: 2 packages finished [3.72s]
```

Both packages have been built. You will have to do that every time you add or modify a node inside a package.

The important thing to notice is this line: `Finished <<< <package_name> [time]`. This means that the package was correctly built. Even if you see additional warning logs, if you also see the `Finished` line, you know the package has been built.

> **Note**
> After you build any package, you also have to source your workspace so that the environment is aware of the new changes. You can do any of the following:
>
> - Open a new terminal as everything is configured in the `.bashrc` file
>
> - Source the `setup.bash` script directly (`source ~/ros2_ws/install/setup.bash`)
>
> - Source the `.bashrc` manually (`source ~/.bashrc`)

Creating a package   65

To build only a specific package, you can use the `--packages-select` option, followed by the name of the package. Here's an example:

```
$ colcon build --packages-select my_py_pkg
Starting >>> my_py_pkg
Finished <<< my_py_pkg [1.01s]
Summary: 1 package finished [1.26s]
```

This way, you don't need to build your entire application every time and can just focus on one package.

Now that we have created some packages and we know how to build them, we can create nodes in the packages. But how are we going to organize them?

How are nodes organized in a package?

To develop a ROS 2 application, you will write code inside nodes. As seen in *Chapter 3*, *node* is simply the name of a ROS 2 program.

A node is a subprogram of your application, responsible for one thing. If you have two different functionalities to implement, then you will have two nodes. Nodes communicate with each other using ROS 2 communications (topics, services, and actions).

You will organize your nodes inside packages. For one package (sub-part of your application), you can have several nodes (functionalities). To fully understand how to organize packages and nodes, you need practice and experience. For now, let's just get an idea with an example.

Let's come back to the package architecture we had in *Figure 4.1*, and add nodes inside the packages:

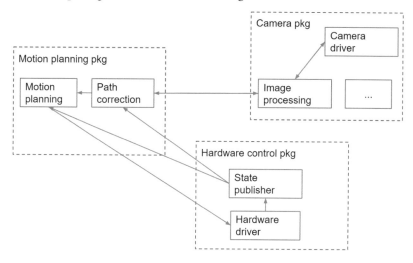

Figure 4.2 – Example of a package organization with nodes

As you can see, in the camera package, we could have one node responsible for handling the camera hardware. This node would send images to an image processing node, and this latter would extract the coordinates of objects for the robot to pick up.

In the meantime, a motion planning node (in the motion planning package) would compute the movements that the robot should perform, given a specific command. A path correction node can support this motion planning using the data received from the image processing node.

Finally, to make the robot move, a hardware driver node would be responsible for hardware communication (motors, encoders) and receive commands from the motion planning node. An additional state publisher node could be here to publish additional data about the robot for other nodes to use.

This node organization is purely fictitious and is here just to give you a general idea of how a ROS 2 application can be designed, and which roles a node can have in this application.

Now, you are (finally) going to write your first ROS 2 node. ROS 2 requires quite a lot of installation and configuration before you can actually write some code, but good news, we have completed all of this and can now focus on the code.

We won't do anything too complicated for now; we won't dive into complex features or communications. We will write a basic node that you can use as a template to start any future node. We will also build the node and see how to run it.

Creating a Python node

Let's create our first Python node, or in other words, our first ROS 2 Python program.

The processes of creating Python and C++ nodes are very different. That's why I have written a separate section for each of them. We will start with Python, with complete step-by-step explanations. Then we will see how to do the same with C++. If you want to follow the C++ node section, make sure to read this one first.

To create a node, you will have to do the following:

1. Create a file for the node.
2. Write the node. We will use **Object Oriented Programming** (**OOP**), as officially recommended for ROS 2 (and almost every existing ROS 2 code you find uses OOP).
3. Build the package in which the node exists.
4. Run the node to test it.

Let's get started with our first Python node.

Creating a file for the node

To write a node, we first need to create a file. Where should we create this file?

If you remember, when we created the my_py_pkg package, another my_py_pkg directory was created inside the package. This is where we will write the node. For every Python package, you have to go to the directory which has the same name as the package. If your package name is abc, then you'll go to ~/ros2_ws/src/abc/abc/.

Create a new file in this directory and make it executable:

```
$ cd ~/ros2_ws/src/my_py_pkg/my_py_pkg/
$ touch my_first_node.py
$ chmod +x my_first_node.py
```

After this, open this file to write in it. You can use any text editor or IDE you want here, as long as you don't get lost in all the files.

If you have no idea what to use, I suggest using VS Code with the ROS extension (as explained in *Chapter 2*). This is the tool I'm using for all ROS development.

> **Note**
>
> If you are using VS Code, the best way to open it is to first navigate to the src directory of your workspace in a terminal, and then open it. This way, you have access to all the packages in your workspace, and it will make things easier with recognized dependencies and auto-completion:
>
> ```
> $ cd ~/ros2_ws/src/
> $ code .
> ```

Writing a minimal ROS 2 Python node

Here is the starting code for any Python node you will create. You can write this code into the my_first_node.py file:

```python
#!/usr/bin/env python3
import rclpy
from rclpy.node import Node

class MyCustomNode(Node):
    def __init__(self):
        super().__init__('my_node_name')

def main(args=None):
    rclpy.init(args=args)
    node = MyCustomNode()
```

```
        rclpy.spin(node)
        rclpy.shutdown()

if __name__ == '__main__':
    main()
```

As you can see, we use OOP here. OOP is everywhere in ROS 2, and this is the default (and recommended) way to write a node.

Let's come back to this code step by step, to understand what it's doing:

```
#!/usr/bin/env python3
import rclpy
from rclpy.node import Node
```

We first import `rclpy`, the Python library for ROS 2. Inside this library, we can get the `Node` class.

We then create a new class that inherits from the `rclpy` `Node` class:

```
class MyCustomNode(Node):
    def __init__(self):
        super().__init__('my_node_name')
```

In this class, make sure you call the parent constructor with `super()`. This is also where you will specify the node name.

This node is not doing anything for now; we will add a few functionalities in a minute. Let's finish the code:

```
def main(args=None):
    rclpy.init(args=args)
    node = MyCustomNode()
    rclpy.spin(node)
    rclpy.shutdown()
```

After the class, we create a `main()` function in which we perform the following actions:

1. Initialize ROS 2 communications with `rclpy.init()`. This should be the first line in your `main()` function.

2. Create an object from the `MyCustomNode` class we wrote before. This will initialize the node. There's no need to destroy the node later, as this will happen automatically when the program exits.

3. Make the node spin. If you omit this line, the node will be created, then the program will exit, and the node will be destroyed. Making the node spin means that we block the execution here, the program stays alive, and thus the node stays alive. In the meantime, as we will see shortly, all registered callbacks for the node can be processed. When you press *Ctrl + C*, the node will stop spinning and this function will return.

4. After the node is killed, shut down ROS 2 communications with `rclpy.shutdown()`. This will be the last line of your `main()` function.

This is how all your ROS 2 programs will work. As you can see, the node is in fact an object that we create within the program (the node is not the program in itself, but still, it is quite common to refer to the word "node" when we talk about the program). After being created, the node can stay alive and play its part while it is spinning. We will come back to this *spinning* shortly.

Finally, we also have added these two lines:

```python
if __name__ == '__main__':
    main()
```

This is a pure Python thing and has nothing to do with ROS 2. It just means that if you run the Python script directly, the `main()` function will be called, so you can try your program without having to install it with `colcon`.

Great, you have written your first minimal Python node. Before you build and run it, add one more line in the Node's constructor, so it can do something:

```python
class MyCustomNode(Node):
    def __init__(self):
        super().__init__('my_node_name')
        self.get_logger().info("Hello World")
```

This line will print `Hello World` when the node starts.

As the `MyCustomNode` class inherits from the `Node` class, we get access to all the ROS 2 functionalities for nodes. This will make things quite convenient for us. Here, you have an example with the logging functionality: we get the `get_logger()` method from `Node`. Then, with the `info()` method, we can print a log with the info level.

Building the node

You are now going to build the node so that you can run it.

You might think: why do we need to build a Python node? Python is an interpreted language; couldn't we just run the file itself?

Yes, this is true: you could test the code just by running it in the terminal (`$ python3 my_first_node.py`). However, what we want to do is actually install the file in our workspace, so we can start the node with `ros2 run`, and later on, from a launch file.

We usually use the word "build", because to install a Python node, we have to run `colcon build`.

To build (install) the node, we need to do one more thing in the package. Open the `setup.py` file from the `my_py_pkg` package. Locate `entry_points` and `'console_scripts'` at the end of the file. For each node we want to build, we have to add one line inside the `'console_scripts'` array:

```
entry_points={
    'console_scripts': [
        "test_node = my_py_pkg.my_first_node:main"
    ],
},
```

Here is the syntax:

```
<executable_name> = <package_name>.<file_name>:<function_name>.
```

There are a few important things to correctly write this line:

- First, choose an executable name. This will be the name you use with `ros2 run <pkg_name> <executable_name>`.
- For the filename, skip the `.py` extension.
- The function name is `main`, as we have created a `main()` function in the code.
- If you want to add another executable for another node, don't forget to add a comma between each executable and place one executable per line.

> **Note**
> When learning ROS 2, there is a common confusion between the node name, filename, and executable name:
>
> - Node name: defined inside the code, in the constructor. This is what you'll see with the `ros2 node list`, or in `rqt_graph`.
>
> - Filename: the file where you write the code.
>
> - Executable name: defined in `setup.py` and used with `ros2 run`.
>
> In this first example, I made sure to use a different name for each so you can be aware that these are three different things. But sometimes all three names could be the same. For example, you could create a `temperature_sensor.py` file, then name your node and your executable `temperature_sensor`.

Now that you have given the instructions to create a new executable, go to your workspace root directory and build the package:

```
$ cd ~/ros2_ws/
$ colcon build
```

You can also add `--packages-select my_py_pkg` to only build this package.

The executable should now be created and installed in the workspace (it will be placed inside the `install` directory). We can say that your Python node has been built, or installed.

Running the node

Now you can run your first node, but just before that, make sure that the workspace is sourced in your environment:

```
$ source ~/.bashrc
```

This file already contains the line to source the workspace; you could also just open a new terminal, or source the `setup.bash` script from the workspace.

You can now run your node using `ros2 run` (if you have any doubts, go back to the experiments we did in *Chapter 3*):

```
$ ros2 run my_py_pkg test_node
[INFO] [1710922181.325254037] [my_node_name]: Hello World
```

Great, we see the log `Hello World`. Your first node is successfully running. Note that we wrote `test_node` in the `ros2 run` command, as it's the executable name we chose in the `setup.py` file.

Now, you might notice that the program is hanging there. The node is still alive because it is spinning. To stop the node, press *Ctrl + C*.

Improving the node – timer and callback

At this point, you might feel that writing, building, and running a node is a long and complicated process. It's actually not that complex, and it gets easier with each new node that you create. On top of that, modifying an existing node is even easier. Let's see that now.

The node we ran is very basic. Let's add one more functionality and do something more interesting.

Our node is printing one piece of text when it's started. We now want to make the node print a string every second, as long as it's alive.

This behavior of "doing X action every Y seconds" is very common in robotics. For example, you could have a node that "reads a temperature every 2 seconds", or that "gives a new motor command every 0.1 seconds".

How to do that? We will add a **timer** to our node. A timer will trigger a **callback** function at a specified rate.

Let's go back to the code and modify the `MyCustomNode` class. The rest of the code stays the same:

```
class MyCustomNode(Node):
    def __init__(self):
        super().__init__('my_node_name')
        self.counter_ = 0
        self.timer_ = self.create_timer(1.0, self.print_hello)
    
    def print_hello(self):
        self.get_logger().info("Hello " + str(self.counter_))
        self.counter_ += 1
```

We still have the constructor with `super()`, but now the log is in a separate method. Also, instead of just printing `Hello World`, here we create a `counter_` attribute that we increment every time we use the log.

> **Note**
>
> If you're wondering why there is a trailing underscore `_` at the end of each class attribute, this is a common OOP convention that I follow to specify that a variable is a class attribute. It's simply a visual help and has no other function. You can follow the same convention or use another one—just make sure to stay consistent within one project.

The most important line is the one to create the timer. To create the timer we use the `create_timer()` method from the `Node` class. We need to give two arguments: the rate at which we want to call the function (float number), and the callback function. Note that the callback function should be specified without any parenthesis.

This instruction means that we want to call the `print_hello` method every `1.0` second.

Let's now try the code. As we have already specified how to create an executable from this file in the `setup.py` file, we don't need to do it again.

All we have to do is to build, source, and run. Remember: "build, source, run." Every time you create a new node, or modify an existing one, you have to "build, source, run."

In a terminal, go to the root directory of your ROS 2 workspace and build the package:

```
$ cd ~/ros2_ws/
$ colcon build --packages-select my_py_pkg
```

> **Note**
>
> On top of `--packages-select <pkg_name>`, you can add the `--symlink-install` option, so you won't have to build the package every time you modify your Python nodes; for example, `$ colcon build --packages-select my_py_pkg --symlink-install`.
>
> You might see some warning logs, but as long as you see the line starting with `Finished <<< my_py_pkg`, it worked correctly. This will install the executable, but then if you modify the code, you should be able to run it without building it again.
>
> Two important things: this only works for Python packages, and you still have to build the package for any new executable you create.

Then, from this terminal or another one, source and run the following:

```
$ source ~/.bashrc
$ ros2 run my_py_pkg test_node
[INFO] [1710999909.533443384] [my_node_name]: Hello 0
[INFO] [1710999910.533169531] [my_node_name]: Hello 1
[INFO] [1710999911.532731467] [my_node_name]: Hello 2
[INFO] [1710999912.534052411] [my_node_name]: Hello 3
```

As you can see, the process of build, source, and run is quite fast and not that complicated. Here, we can see that the node prints a log every second, and the counter increments in each new log.

Now, how is this possible? How is the `print_hello()` method called? We have created a timer, yes, but nowhere in the code have we actually called `print_hello()` directly.

It works because the node is spinning, thanks to `rclpy.spin(node)`. This means that the node is kept alive, and all registered callbacks can be called during this time. What we do with `create_timer()` is simply to register a callback, which can then be called when the node is spinning.

This was your first example of a callback, and as you will see in the following chapters of the book, everything runs with callbacks in ROS 2. At this point, if you still have some trouble with the syntax, the callbacks, and the spinning, don't worry too much. As you make progress with the book, you will repeat this process many times. When learning ROS 2, understanding comes with hands-on experience.

We are now done with this Python node. With what you've seen here, you should be able to create your own new Python nodes (in the same package or another package). Let's now switch to C++. If you are only interested in learning ROS 2 with Python for now, you can skip the C++ section.

Creating a C++ node

We are going to do exactly the same thing we did for the Python node: create a file, write the node, build, source, and run.

Make sure you have read the previous Python section as I will not repeat everything here. We will basically just see how to apply the process for a C++ node.

To create a C++ node, we first need a C++ package. We will use the my_cpp_pkg package that we created previously.

Writing a C++ node

Let's create a file for the node. Go to the src directory inside the my_cpp_pkg package and create a .cpp file:

```
$ cd ~/ros2_ws/src/my_cpp_pkg/src/
$ touch my_first_node.cpp
```

You could also create the file directly from your IDE and not use the terminal.

Now, if you haven't done this previously, open your workspace with VS Code or any other IDE:

```
$ cd ~/ros2_ws/src/
$ code .
```

Open my_first_node.cpp. Here is the minimal code to write a C++ node:

```cpp
#include "rclcpp/rclcpp.hpp"

class MyCustomNode : public rclcpp::Node
{
public:
    MyCustomNode() : Node("my_node_name")
    {
    }
private:
};

int main(int argc, char **argv)
{
    rclcpp::init(argc, argv);
    auto node = std::make_shared<MyCustomNode>();
    rclcpp::spin(node);
    rclcpp::shutdown();
    return 0;
}
```

> **Note**
>
> If you are using VS Code and you type this code, you might see an include error for the `rclcpp` library. Make sure to save the file and wait a few seconds. If the include is still not recognized, go to the **Extensions** tab and disable and re-enable the ROS extension.

As you can see (and this was similar with Python), in ROS 2 we heavily use OOP with C++ nodes.

Let's analyze this code step by step:

```
#include "rclcpp/rclcpp.hpp"
```

We first include `rclcpp`, the C++ library for ROS 2. This library contains the `rclcpp::Node` class:

```
class MyCustomNode : public rclcpp::Node
{
public:
    MyCustomNode() : Node("my_node_name")
    {
    }
private:
};
```

As we did for Python, we have created a class that inherits from the Node class. The syntax is different, but the principle is the same. From this Node class, we will be able to access all the ROS 2 functionalities: logger, timer, and so on. As you can see, we also specify the node name in the constructor. For now, the node does nothing; we will add more functionalities in a minute:

```
int main(int argc, char **argv)
{
    rclcpp::init(argc, argv);
    auto node = std::make_shared<MyCustomNode>();
    rclcpp::spin(node);
    rclcpp::shutdown();
    return 0;
}
```

You need a `main()` function if you want to be able to run your C++ program. In this function, we do exactly the same thing as for Python, with just some differences in the syntax:

1. Initialize ROS 2 communications with `rclcpp::init()`.
2. Create a node object from your newly written class. As you can see, we don't create an object directly, but a shared pointer to that object. In ROS 2 and C++, almost everything you create will be a smart pointer (shared, unique, and so on).

3. We then make the node spin with `rclcpp::spin()`.
4. Finally, when the node is stopped (*Ctrl + C*), we shut down all ROS 2 communications with `rclcpp::shutdown()`.

This structure for the `main()` function will be very similar for all your ROS 2 programs. As you can see, once again, the node is not the program in itself. The node is created inside the program.

Before we go further and build, source, and run our node, let's improve it now with a timer, a callback, and a log.

Modify the `MyCustomNode` class, and leave the rest as it is:

```cpp
class MyCustomNode : public rclcpp::Node
{
public:
    MyCustomNode() : Node("my_node_name"), counter_(0)
    {
        timer_ = this->create_wall_timer(std::chrono::seconds(1),
    std::bind(&MyCustomNode::print_hello, this));
    }

    void print_hello()
    {
        RCLCPP_INFO(this->get_logger(), "Hello %d", counter_);
        counter_++;
    }
private:
    int counter_;
    rclcpp::TimerBase::SharedPtr timer_;
};
```

This code example will do the same thing as for the Python node. We create a timer so that we can call a callback function every `1.0` second. In this callback function, we print `Hello` followed by a counter that we increment every time.

There are some specificities related to C++:

- For the timer, we have to create a class attribute. As you can see we also create a shared pointer here: `rclcpp::TimerBase::SharedPtr`.
- We use `this->create_wall_timer()` to create the timer. `this->` is not required here, but I have added it to emphasize that we are using the `create_wall_timer()` method from the `Node` class.
- To specify the callback in the timer, as we are in a C++ class, we have to use `std::bind(&ClassName::method_name, this)`. Make sure you don't use any parenthesis for the method name.

Building and running the node

We can't just run the C++ file; we first have to compile it and create an executable. To do this, we will edit the CMakeLists.txt file. Open this file, and after a few lines, you will find something like this:

```
# find dependencies
find_package(ament_cmake REQUIRED)
find_package(rclcpp REQUIRED)
```

The line to find rclcpp is here because we provided --dependencies rclcpp when we created the package with ros2 pkg create. Later on, if your nodes in this package require more dependencies, you can add the dependencies here, one per line.

Just after this line, add an extra new line, and then the following instructions:

```
add_executable(test_node src/my_first_node.cpp)
ament_target_dependencies(test_node rclcpp)

install(TARGETS
  test_node
  DESTINATION lib/${PROJECT_NAME}/
)
```

To build a C++ node, we need to do three things:

1. Add a new executable with the add_executable() function. Here, you have to choose a name for the executable (the one that will be used with ros2 run <pkg_name> <executable_name>), and we also have to specify the relative path to the C++ file.
2. Link all dependencies for this executable with the ament_target_dependencies() function.
3. Install the executable with the install() instruction, so that we can find it when we use ros2 run. Here, we put the executable in a lib/<package_name> directory.

Then, for each new executable you create, you need to repeat *steps 1* and *2* and add the executable inside the install() instruction, one per line without any commas. There's no need to create a new install() instruction for each executable.

> **Note**
> The end of your CMakeLists.txt will contain a block starting with if(BUILD_TESTING), and then ament_package(). As we are not doing any build testing here, you can remove the entire if block. Just make sure to keep the ament_package() line, which should be the last line of the file.

Writing and Building a ROS 2 Node

You can now build the package with `colcon build`, which is going to create and install the executable:

```
$ cd ~/ros2_ws/
$ colcon build --packages-select my_cpp_pkg
```

If you get any error during the build process, make sure to fix your code first, and then build again. Then, you can source your environment, and run your executable:

```
$ source ~/.bashrc
$ ros2 run my_cpp_pkg test_node
[INFO] [1711006463.017149024] [my_node_name]: Hello 0
[INFO] [1711006464.018055674] [my_node_name]: Hello 1
[INFO] [1711006465.015927319] [my_node_name]: Hello 2
[INFO] [1711006466.015355747] [my_node_name]: Hello 3
```

As you can see, we run the `test_node` executable (built from `my_first_node.cpp` file), which is going to start the `my_node_name` node.

You have now successfully written a C++ node. For each new node that you create, you will have to create a new C++ file, write the node class, set the build instructions for a new executable in `CMakeLists.txt`, and build the package. Then, to start the node, source the environment and run the executable with `ros2 run`.

Node template for Python and C++ nodes

All the nodes we start in this book will follow the same structure. As additional help to get started quickly, I have created a node template you can use to write the base of any Python or C++ node. I use these templates myself when creating new nodes, as the code can be quite repetitive.

You can copy and paste the templates either from this book directly, or download them from the GitHub repository: `https://github.com/PacktPublishing/ROS-2-from-Scratch`.

Template for a Python node

Use this code to start any new Python node:

```
#!/usr/bin/env python3
import rclpy
from rclpy.node import Node

class MyCustomNode(Node): # MODIFY NAME
    def __init__(self):
        super().__init__("node_name") # MODIFY NAME

def main(args=None):
```

```python
    rclpy.init(args=args)
    node = MyCustomNode() # MODIFY NAME
    rclpy.spin(node)
    rclpy.shutdown()

if __name__ == "__main__":
    main()
```

All you have to do is remove the `MODIFY NAME` comments and change the class name (`MyCustomNode`) and the node name (`"node_name"`). It's better to use names that make sense. For example, if you are writing a node to read data from a temperature sensor, you could name the class `TemperatureSensorNode`, and the node could be `temperature_sensor`.

Template for a C++ node

Use this code to start any new C++ node:

```cpp
#include "rclcpp/rclcpp.hpp"

class MyCustomNode : public rclcpp::Node // MODIFY NAME
{
public:
    MyCustomNode() : Node("node_name") // MODIFY NAME
    {
    }

private:
};

int main(int argc, char **argv)
{
    rclcpp::init(argc, argv);
    auto node = std::make_shared<MyCustomNode>(); // MODIFY NAME
    rclcpp::spin(node);
    rclcpp::shutdown();
    return 0;
}
```

Remove the `MODIFY NAME` comments and rename the class and the node.

Those two templates will allow you to start your nodes more quickly. I recommend you to use them as much as you can.

Introspecting your nodes

To finish this chapter, we will practice a bit more with the `ros2 node` command line.

So far, you have seen how to write a node, build it, and run it. One missing part is to know how to introspect your nodes. Even if a node can run, it doesn't mean it will do exactly what you want it to do.

Being able to introspect your nodes will help you fix errors that you might have made in your code. It will also allow you to easily find more information about other nodes that you are starting but didn't write (as we did in the discovery phase in *Chapter 3*).

For each core concept in *Part 2*, we will take a bit of time to experiment with the command-line tools related to the concept. The command-line tool for nodes is `ros2 node`.

First, and before we use `ros2 node`, we have to start a node. As a recap, to start a node, we use `ros2 run <package_name> <executable_name>`. If we start the Python node we have created in this chapter, we use this:

```
$ ros2 run my_py_pkg test_node
```

Only after we have started a node can we do some introspection with `ros2 node`.

ros2 node command line

To list all running nodes, use `ros2 node list`:

```
$ ros2 node list
/my_node_name
```

We find the name of the node, which we defined in the code.

Once we have the node name, we can get more info about it with `ros2 node info <node_name>`:

```
$ ros2 node info /my_node_name
/my_node_name
  Subscribers:
  Publishers:
    /parameter_events: rcl_interfaces/msg/ParameterEvent
    /rosout: rcl_interfaces/msg/Log
  Service Servers:
    /my_node_name/describe_parameters: rcl_interfaces/srv/DescribeParameters
    /my_node_name/get_parameter_types: rcl_interfaces/srv/GetParameterTypes
    /my_node_name/get_parameters: rcl_interfaces/srv/GetParameters
    /my_node_name/get_type_description: type_description_interfaces/srv/GetTypeDescription
    /my_node_name/list_parameters: rcl_interfaces/srv/ListParameters
```

```
    /my_node_name/set_parameters: rcl_interfaces/srv/SetParameters
    /my_node_name/set_parameters_atomically: rcl_interfaces/srv/
SetParametersAtomically
  Service Clients:
  Action Servers:
  Action Clients:
```

As you can see, there are quite a lot of things on the terminal. We will get to know all of them in the following chapters. With `ros2 node info <node_name>` you can see all topics (publishers/subscribers), services, and actions running for this node.

Changing the node name at run time

As we progress throughout the book, I will give you additional tips for working with ROS 2 and the command line. Here is one: when starting an executable, you can choose to use the default node name (the one defined in the code) or replace it with a new name.

To add any additional argument to `ros2 run`, first add `--ros-args` (only once).

Then, to rename the node, add `-r __node:=<new_name>`. `-r` means remap; you could also use `--remap`. For example, if we want to name the node `abc`, we could use this:

```
$ ros2 run my_py_pkg test_node --ros-args -r __node:=abc
[INFO] [1711010078.801996629] [abc]: Hello 0
[INFO] [1711010079.805748394] [abc]: Hello 1
```

As you can see from the logs, instead of `my_node_name`, we see `abc`.

List all running nodes:

```
$ ros2 node list
/abc
```

This functionality can be very helpful and gives you more control over how to start a node, without having to modify the code directly.

> **Note**
> When running multiple nodes, you should make sure that each node has a unique name. Having two nodes with the same name can lead to some unexpected issues that can take a long time to debug. In the future, you will see that you may want to run the same node several times, for example, three `temperature_sensor` nodes, one each for a different sensor. You could rename them so that you have `temperature_sensor_1`, `temperature_sensor_2`, and `temperature_sensor_3`.

Summary

In this chapter, you have created your first node. Let's do a quick recap of all the steps.

Before creating any node, you need to follow these steps:

1. You first need to create and set up a ROS 2 workspace.
2. In this workspace, you can create several packages (Python or C++) that represent different sub-parts of your application.

Then, in one package you can create one or several nodes. For each node, you will have to do the following:

1. Create a file inside the package.
2. Write the node (using the OOP template as a base).
3. Set the build instructions (`setup.py` for Python, `CMakeLists.txt` for C++).
4. Build the package.

To run the node, don't forget to source the workspace first, and then start the node with `ros2 run <pkg_name> <executable_name>`.

Finally, you can introspect your nodes and even change their names when you start them, using the `ros2 node` command line.

Feel free to come back to this chapter anytime to see the complete process of creating a node for both Python and C++. All the code is available on GitHub at `https://github.com/PacktPublishing/ROS-2-from-Scratch`. There you can find the OOP template code for Python and C++, `my_py_pkg` package, and `my_cpp_pkg` package.

In this chapter, you have also seen how to create a timer and a callback function. You have a better idea of how the spin mechanism works, and how it allows the node to stay alive and run the callbacks. This will be very useful for the following chapters.

In the next chapter, we will see how nodes communicate with each other using topics. You will write your own topics (publishers/subscribers) inside nodes and experiment with them.

5
Topics – Sending and Receiving Messages between Nodes

Now that you can write nodes, how can you make several nodes communicate with each other, and how can you interact with existing nodes in an application?

There are three kinds of communication in ROS 2: topics, services, and actions. In this chapter, we will dive into ROS 2 topics.

To understand how topics work, we will start with a real-life analogy. This will allow you to grasp the concept using existing and common knowledge. Then, you will dive into the code and write a publisher and a subscriber inside a node—first with existing interfaces, and then by building custom interfaces. You will also use ROS 2 tools such as the `ros2` command line and `rqt_graph` to introspect topics and unlock more functionalities.

By the end of this chapter, you will be able to make your nodes communicate with each other using ROS 2 topics. You will learn by writing code and will be provided with an additional challenge at the end of this chapter.

Topics are used everywhere in ROS 2. Whether you wish to create an application from scratch or use existing ROS plugins, you will have to use topics.

We will use the code inside the `ch4` folder in this book's GitHub repository (`https://github.com/PacktPublishing/ROS-2-from-Scratch`) as a starting point. You can find the final code in the `ch5` folder.

In this chapter, we will cover the following topics:

- What is a ROS 2 topic?
- Writing a topic publisher
- Writing a topic subscriber

- Additional tools to handle topics
- Creating a custom interface for a topic
- Topic challenge – closed-loop control

What is a ROS 2 topic?

You discovered the concept of topics through hands-on experiments in *Chapter 3*. With this, you should have a basic intuition of how things work.

I am now going to start from scratch again and explain topics—not by running code, but by using a real-life analogy that makes it easier to understand. We will build an example, step by step, and then recap the most important points.

A publisher and a subscriber

For this analogy, I will use radio transmitters and receivers. As this is a simplified example, not everything I'll say about radio will be correct, but the point here is to understand ROS 2 topics.

Let's start with one radio transmitter. This radio transmitter will send some data at a given frequency. To make it easier for people to remember, this frequency is usually represented by a number, such as *98.7*. We can even think of *98.7* as a name. If you want to listen to the radio, you know you have to connect your device to *98.7*.

In this case, we can say that *98.7* is a topic. The radio transmitter is a **publisher** on this topic:

Figure 5.1 – Radio transmitter publishing to the 98.7 topic

Now, let's say you want to listen to that radio from your phone. You will ask your phone to connect to *98.7* to receive the data.

With this analogy, the phone is then a **subscriber** to the *98.7* topic.

One important thing to note here is that both the radio transmitter and the phone must use the same type of frequency. For example, if the radio transmitter is using an AM signal, and if the phone is trying to decode an FM signal, it will not work.

Similarly, with ROS 2 topics, both the publisher and subscriber must use the same data type. This data type is called an **interface**.

This is what defines a topic: a **name** and an interface:

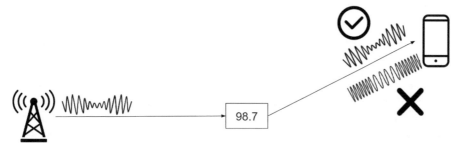

Figure 5.2 – Publisher and subscriber using the same interface

With that, the communication is complete. The radio transmitter publishes an AM signal on the *98.7* topic. The phone subscribes to the *98.7* topic, decoding an AM signal.

Multiple publishers and subscribers

In real life, there won't be just one device trying to listen to the radio. Let's add a few more devices, each one subscribing to the *98.7* topic and decoding an AM signal:

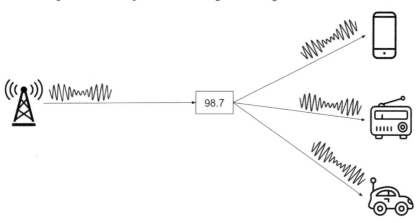

Figure 5.3 – Topic with multiple subscribers

As you can see, one topic can have several subscribers. Each subscriber will get the same data. On the other hand, we could also have several publishers for one topic.

Imagine that there is another radio transmitter, also publishing an AM signal to *98.7*. In this case, both the data from the first transmitter and the second transmitter are received by all listening devices:

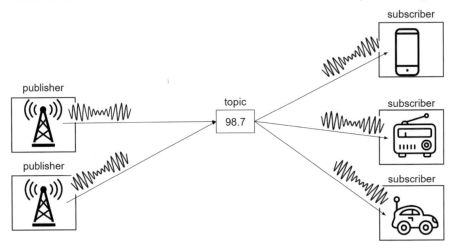

Figure 5.4 – Multiple publishers and subscribers

The preceding figure shows boxes. Each box represents a node. Thus, we have two radio transmitter nodes, both containing a publisher to the *98.7* topic. We also have three nodes (phone, radio receiver, and car), each one containing a subscriber to *98.7*.

Note that one subscriber is not aware of the other subscribers. When you listen to the radio on your phone, you have no idea who else is listening to the radio, and on what device.

Also, the phone, the radio receiver and the car are not aware of who is publishing on the radio. They only know they have to subscribe to *98.7*; they don't know what's behind it.

On the other side, both radio transmitters are not aware of each other and of who is receiving the data. They just publish on the topic, regardless of who is listening. Thus, we say that topics are **anonymous**. Publishers and subscribers are not aware of other publishers and subscribers. They only publish or subscribe to a topic, using its name and interface.

Any combination of publishers and subscribers is possible. For example, you could have two publishers on the topic and zero subscribers. In this case, the data is still correctly published, but no one receives it. Alternatively, you could have zero publishers and one or more subscribers. The subscribers will listen to the topic but will receive nothing.

Multiple publishers and subscribers inside one node

A node is not limited to having just one publisher or one subscriber.

Let's add another radio to our example. We will name it *101.3*, and its data type is FM signal.

The second radio transmitter is now publishing both on the *98.7* topic and the *101.3* topic, sending the appropriate type of data for each topic. Let's also make the car listen to the *101.3* topic:

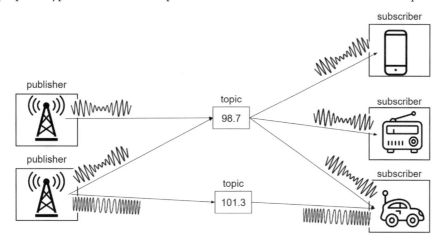

Figure 5.5 – A node with two publishers

As you can see, the second radio transmitter can publish on several topics, so long as it uses the correct name and interface for each topic.

Now, imagine that the car, while listening to the radio, is also sending its GPS coordinates to a remote server. We could create a topic named `car_location`, and the interface would contain a latitude and a longitude. The car node now contains one subscriber to the *98.7* topic, and one publisher to the `car_location` topic:

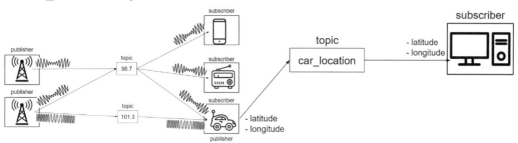

Figure 5.6 – A node with both a publisher and a subscriber

In the preceding figure, I have also added another node for the server, represented by a computer. The server node will subscribe to the `car_location` topic so that it can receive the GPS coordinates. Of course, both the publisher and subscriber are using the same interface (latitude and longitude).

Thus, inside one node, you can have any number of publishers and subscribers to different topics with different data types. A node can communicate with several nodes at the same time.

Wrapping things up

ROS 2 nodes can send messages to other nodes using topics.

Topics are mostly used to send data streams. For example, you could create a hardware driver for a camera sensor, and publish images taken from the camera. Other nodes can then subscribe to the topic and receive the images. You could also publish a stream of commands for a robot to move, and so on.

There are many possibilities for when to use topics, and you will get to know more about them as you progress throughout this book.

Here are some important points about how topics work:

- A topic is defined by a name and an interface.
- A topic name must start with a letter and can be followed by other letters, numbers, underscores, tildes, and slashes. For the real-life analogy with radio, I used numbers with dots as topic names. Although it made the examples easier, this is not valid for ROS 2 topics. To make it valid, instead of *98.7*, we would have to create a topic named `radio_98_7`.
- Any publisher or subscriber to a topic must use the same interface.
- Publishers and subscribers are anonymous. They are not aware of each other; they just know they are publishing or subscribing to a topic.
- A node can contain several publishers and subscribers to different topics.

Now, how do you create a publisher or a subscriber?

You will do this by adding some code to your nodes. As you saw previously, you can write a Python node using `rclpy` and a C++ node using `rclcpp`. With those two libraries, you can create publishers and subscribers directly in your nodes.

Writing a topic publisher

In this section, you'll write your first ROS 2 publisher. To work on the core concepts, we will create a new ROS 2 application and build upon it in the following chapters. This application will be super minimalistic so that we can focus on the concept we want to learn, nothing else.

What we want to do for now is publish a number on a topic. This topic is new and we will *create* it. You don't really create a topic—you create a publisher or a subscriber to that topic. This will automatically create the topic name, which will be registered on the graph.

To write a publisher, we need a node. We could use the first node we created in the previous chapter, but the purpose of the node is not the same. Hence, we will create a new node named `number_publisher`. In this node, we will create a publisher. As to the topic we want to publish to, we will have to choose a name and an interface.

Now, let's get started with Python.

Writing a Python publisher

To write a publisher, we need to create a node; to create a node, we need a package. To make things simple, let's continue using the `my_py_pkg` package.

Creating a node

Navigate inside the `my_py_pkg` package, create a Python file, and make it executable:

```
$ cd ~/ros2_ws/src/my_py_pkg/my_py_pkg/
$ touch number_publisher.py
$ chmod +x number_publisher.py
```

Now, open this file, use the node OOP template (given in *Chapter 4*), and modify the required fields to give names that make sense:

```python
#!/usr/bin/env python3
import rclpy
from rclpy.node import Node

class NumberPublisherNode(Node):
    def __init__(self):
        super().__init__("number_publisher")

def main(args=None):
    rclpy.init(args=args)
    node = NumberPublisherNode()
    rclpy.spin(node)
    rclpy.shutdown()

if __name__ == "__main__":
    main()
```

Now that you have a `main()` function and a `NumberPublisherNode` class for your node, we can create a publisher.

Adding a publisher to the node

Where can we create a publisher in this node? We will do that in the constructor.

And before we write the code, we need to ask ourselves a question: what is the name and the interface for this topic?

- **Case 1**: You're publishing to a topic that already exists (other publishers or subscribers on that topic), and then you use the same name and interface
- **Case 2**: You create a publisher for a new topic (what we are doing now), and then you have to choose a name and interface

For the name, let's keep it simple and use `number`. If we publish a number, we can expect to receive this number on a `number` topic. If you were to publish a temperature, you could name the topic `temperature`.

For the interface, you have two choices: use an existing interface or create a custom one. To get started, we will use an existing interface. To make this easier, I will just tell you what to use; you'll learn how to find other interfaces yourself later.

Let's use `example_interfaces/msg/Int64`. To get more details about what's in the interface, we can run `ros2 interface show <interface_name>` in the Terminal:

```
$ ros2 interface show example_interfaces/msg/Int64
# Some comments
int64 data
```

Great—this is exactly what we need: an `int64` number.

Now that we have this information, let's create the publisher. First, import the interface, and then create the publisher in the constructor:

```
import rclpy
from rclpy.node import Node
from example_interfaces.msg import Int64

class NumberPublisherNode(Node):
    def __init__(self):
        super().__init__("number_publisher")
        self.number_publisher_ = self.create_publisher(Int64, "number", 10)
```

To import the interface, we must specify the name of the package (`example_interfaces`), then the folder name for topic messages (`msg`), and finally the class for the interface (`Int64`).

To create the publisher, we must use the `create_publisher()` method from the `Node` class. Inheriting from this class gives us access to all ROS 2 functionalities. In this method, you have to provide three arguments:

- **Topic interface**: We'll use `Int64` from the `example_interfaces` package.
- **Topic name**: As defined previously, this is `number`.
- **Queue size**: If the messages are published too fast and subscribers can't keep up, messages will be buffered (up to 10 here) so that they're not lost. This can be important if you send large messages (images) at a high frequency, on a lossy network. As we get started, there's no need to worry about this; I recommend that you just set the queue size to `10` every time.

With this, we now have a publisher on the `number` topic. However, if you just run your code like this, nothing will happen. A publisher won't publish automatically on a topic. You have to write the code for that to happen.

Publishing with a timer

A common behavior in robotics is to do *X* action every *Y* seconds—for example, publish an image from a camera every `0.5` seconds, or in this case, publish a number on a topic every `1.0` second. As seen in *Chapter 4*, to do this, you must implement a timer and a callback function.

Modify the code inside the node so that you publish on the topic from a timer callback:

```
def __init__(self):
    super().__init__("number_publisher")
    self.number_ = 2
    self.number_publisher_ = self.create_publisher(Int64, "number", 10)
    self.number_timer_ = self.create_timer(1.0, self.publish_number)
    self.get_logger().info("Number publisher has been started.")

def publish_number(self):
    msg = Int64()
    msg.data = self.number_
    self.number_publisher_.publish(msg)
```

After creating the publisher with `self.create_publisher()`, we create a timer with `self.create_timer()`. Here, we say that we want the `publish_number()` method to be called every `1.0` second. This will happen when the node is spinning.

On top of that, I also added a log at the end of the constructor to say that the node has been started. I usually do this as a best practice so that I can see when the node is fully initialized on the Terminal.

In the `publish_number()` method, we publish on the topic:

1. We create an object from the `Int64` class. This is the interface—in other words, the message to send.
2. This object contains a `data` field. How do we know this? We found this previously when we ran `ros2 interface show example_interfaces/msg/Int64`. Thus, we provide a number in the `data` field of the message. For simplicity, we specify the same number every time we run the callback function.
3. We publish the message using the `publish()` method from the publisher.

This code structure is super common in ROS 2. Any time you want to publish data from a sensor, you will write something similar.

Building the publisher

To try your code, you need to install the node.

Before we do this, since we're using a new dependency (`example_interfaces` package), we also need to add one line to the `package.xml` file of the `my_py_pkg` package:

```
<depend>rclpy</depend>
<depend>example_interfaces</depend>
```

As you add more functionalities inside your package, you will add any other ROS 2 dependency here.

To install the node, open the `setup.py` file from the `my_py_pkg` package and add a new line to create another executable:

```
entry_points={
    'console_scripts': [
        "test_node = my_py_pkg.my_first_node:main",
        "number_publisher = my_py_pkg.number_publisher:main"
    ],
},
```

Make sure you add a comma between each line; otherwise, you could encounter some strange errors when building the package.

Here, we've created a new executable named `number_publisher`.

> **Note**
> This time, as you can see from this example, the node name, filename, and executable name are the same: `number_publisher`. This is a common thing to do. Just remember that those names represent three different things.

Now, go to your workspace root directory and build the my_py_pkg package:

```
$ cd ~/ros2_ws/
$ colcon build --packages-select my_py_pkg
```

You can add --symlink-install if you want to, so that you don't need to run colcon build every time you modify the number_publisher node.

Running the publisher

After the package has been built successfully, source your workspace and start the node:

```
$ source install/setup.bash # or source ~/.bashrc
$ ros2 run my_py_pkg number_publisher
[INFO] [1711526444.403883187] [number_publisher]: Number publisher has been started.
```

The node is running, but apart from the initial log, nothing is displayed. That's normal—we didn't ask the node to print anything else.

How do we know that the publisher is working? We could write a subscriber node right away and see if we receive the messages. But before we do that, we can test the publisher directly from the Terminal.

Open a new Terminal window and list all topics:

```
$ ros2 topic list
/number
/parameter_events
/rosout
```

Here, you can find the /number topic.

> **Note**
>
> As you can see, there is an added leading slash in front of the topic name. We only wrote number in the code, not /number. This is because ROS 2 names (nodes, topics, and so on) are organized inside namespaces. Later, we will see that you can add a namespace to put all your topics or nodes inside the /abc namespace, for example. In this case, the topic name would be /abc/number. Here, as no namespace is provided, a leading slash is added to the name, even if we don't provide it in the code. We could call this the *global* namespace.

With the `ros2 topic echo <topic_name>` command, you can subscribe to the topic directly from the subscriber and see what's being published. We will learn more about this command later in this chapter:

```
$ ros2 topic echo /number
data: 2
---
data: 2
---
```

As you can see, we get one new message per second, which contains a `data` field with a value of 2. This is exactly what we wanted to do in the code.

With that, we've finished our first Python publisher. Let's switch to C++.

Writing a C++ publisher

Here, the process is the same as for Python. We will create a new node, and in this node, add a publisher and a timer. In the timer callback function, we will create a message and publish it.

I will go a bit more quickly in this section as the explanations are the same. We will just focus on the specificities of the C++ syntax with ROS 2.

> **Note**
> For everything related to C++ in this book, make sure you follow the explanations using the GitHub code on the side. I may not provide the full code, only the important snippets that are crucial for comprehension.

Creating a node with a publisher and a timer

First, let's create a new file for our `number_publisher` node in the `my_cpp_pkg` package:

```
$ cd ~/ros2_ws/src/my_cpp_pkg/src/
$ touch number_publisher.cpp
```

Open this file and write the code for the node. You can start from the OOP template and add the publisher, timer, and callback function. The complete code for this chapter can be found in this book's GitHub repository: https://github.com/PacktPublishing/ROS-2-from-Scratch.

I will now comment on a few important lines:

```
#include "rclcpp/rclcpp.hpp"
#include "example_interfaces/msg/int64.hpp"
```

To include an interface for a topic, use `"<package_name>/msg/<message_name>.hpp"`.

Then, in the constructor, add the following:

```
number_publisher_ = this->create_publisher<example_
interfaces::msg::Int64>("number", 10);
```

In C++, we also use the `create_publisher()` method from the `Node` class. The syntax is a bit different since templates are used, but you can still find the topic interface, topic name, and queue size (as a reminder, you can set it to `10` every time).

The publisher is also declared as a private attribute in the class:

```
rclcpp::Publisher<example_interfaces::msg::Int64>::SharedPtr number_
publisher_;
```

As you can see, we use the `rclcpp::Publisher` class, and as for many things in ROS 2, we use a shared pointer. For several common classes, ROS 2 provides `::SharedPtr`, which would be the same thing as writing `std::shared_ptr<the publisher>`.

Let's go back to the constructor:

```
number_timer_ = this->create_wall_timer(std::chrono::seconds(1),
std::bind(&NumberPublisherNode::publishNumber, this));
RCLCPP_INFO(this->get_logger(), "Number publisher has been started.");
```

After creating the publisher, we create a timer to call the `publishNumber` method every `1.0` second. Finally, we print a log so that we know that the constructor code has been executed:

```
void publishNumber()
{
    auto msg = example_interfaces::msg::Int64();
    msg.data = number_;
    number_publisher_->publish(msg);
}
```

This is the callback method. As for Python, we create an object from the interface class, after which we fill any field from this interface and publish the message.

Building and running the publisher

Once you've written the node with the publisher, timer, and callback function, it's time to build it.

As we did for Python, open the `package.xml` file of the `my_cpp_pkg` package and add one line for the dependency to `example_interfaces`:

```
<depend>rclcpp</depend>
<depend>example_interfaces</depend>
```

Then, open the `CMakeLists.txt` file from the `my_cpp_pkg` package and add the following lines:

```
find_package(rclcpp REQUIRED)
find_package(example_interfaces REQUIRED)
add_executable(test_node src/my_first_node.cpp)
ament_target_dependencies(test_node rclcpp)
add_executable(number_publisher src/number_publisher.cpp)
ament_target_dependencies(number_publisher rclcpp example_interfaces)
install(TARGETS
  test_node
  number_publisher
  DESTINATION lib/${PROJECT_NAME}/
)
```

For any new dependency, we need to add a new `find_package()` line.

Then, we create a new executable. Note that we also provide `example_interfaces` in the arguments of `ament_target_dependencies()`. If you omit this, you will get an error during compilation.

Finally, there's no need to re-create the `install()` block. Just add the executable in a new line, without any commas between the lines.

Now, you can build, source, and run:

```
$ cd ~/ros2_ws/
$ colcon build --packages-select my_cpp_pkg
$ source install/setup.bash
$ ros2 run my_cpp_pkg number_publisher
[INFO] [1711528108.225880935] [number_publisher]: Number publisher has been started.
```

The node containing the publisher is up and running. By using `ros2 topic list` and `ros2 topic echo <topic_name>`, you can find the topic and see what's being published.

Now that you've created a publisher and you know it's working, it's time to learn how to create a subscriber for that topic.

Writing a topic subscriber

To continue improving our application, let's create a new node that will subscribe to the `/number` topic. Each number that's received will be added to a counter. We want to print the counter every time it's updated.

As we did previously, let's start the full explanations with Python, and then see the syntax specificities with C++.

Writing a Python subscriber

You can find the complete code for this Python node on GitHub. Many things we need to do here are identical to what we did previously, so I won't fully detail every step. Instead, we will focus on the most important things so that we can write the subscriber.

Creating a Python node with a subscriber

Create a new node named number_counter inside the my_py_pkg package:

```
$ cd ~/ros2_ws/src/my_py_pkg/my_py_pkg/
$ touch number_counter.py
$ chmod +x number_counter.py
```

In this file, you can write the code for the node and add a subscriber. Here's the explanation, step by step:

```
#!/usr/bin/env python3
import rclpy
from rclpy.node import Node
from example_interfaces.msg import Int64
```

Since we want to create a subscriber to receive what we sent with the publisher, we need to use the same interface. Hence, we import Int64 as well. Then, we can create the subscriber:

```
class NumberCounterNode(Node):
    def __init__(self):
        super().__init__("number_counter")
        self.counter_ = 0
        self.number_subscriber_ = self.create_subscription(Int64,
 "number", self.callback_number, 10)
        self.get_logger().info("Number Counter has been started.")
```

As for publishers, we will create subscribers in the node's constructor. Here, we use the create_ subscription() method from the Node class. With this method, you need to provide four arguments:

- **Topic interface**: Int64. This needs to be the same for both the publisher and subscriber.
- **Topic name**: number. This is the same name as for the publisher. Note that I don't provide any additional slash here. This will be added automatically, so the topic name will become /number.
- **Callback function**: Do you remember when I told you that almost everything is a callback in ROS 2? We use a callback method for the subscriber here as well. When the node is spinning, it will stay alive and all registered callbacks will be ready to be called. Whenever a message is published on the /number topic, it will be received here, and we will be able to use it and process it inside the callback method (that we need to implement).
- **Queue size**: As seen previously, you can set it to 10 and forget about it for now.

Now, let's see the implementation for the callback method, which I named `callback_number`:

> **Note**
>
> As a best practice, I recommend naming callback methods for topics `callback_<topic>`. By adding the `callback_` prefix, you make it clear that this method is a callback and shouldn't be called directly in the code. This can prevent lots of errors in the future.

```
def callback_number(self, msg: Int64):
    self.counter_ += msg.data
    self.get_logger().info("Counter:  " + str(self.counter_))
```

In a subscriber callback, you receive the message directly in the parameters of the function. Since we know that `Int64` contains a `data` field, we can access it with `msg.data`.

Now, we add the received number to a `counter_` attribute and print the counter every time with a ROS 2 log.

> **Note**
>
> As a best practice, I have specified the `Int64` type for the `msg` argument of the method. This isn't mandatory for Python code to work, but it adds an extra level of safety (we are sure that we should receive `Int64` and nothing else) and it can sometimes make your IDE work better with auto-completion.

To finish the node, don't forget to add the `main()` function after the `NumberCounterNode` class.

Running the Python subscriber

Now, to try the code, add a new executable to the `setup.py` file of your Python package:

```
entry_points={
    'console_scripts': [
        "test_node = my_py_pkg.my_first_node:main",
        "number_publisher = my_py_pkg.number_publisher:main",
        "number_counter = my_py_pkg.number_counter:main"
    ],
},
```

Then, build the package and source the workspace (from now on, I will not write those commands every time since they're always the same).

Now, run each node (`number_publisher` and `number_counter`) in a different Terminal:

```
$ ros2 run my_py_pkg number_publisher
[INFO] [1711529824.816514561] [number_publisher]: Number publisher has
been started.
$ ros2 run my_py_pkg number_counter
[INFO] [1711528797.363370081] [number_counter]: Number Counter has
been started.
[INFO] [1711528815.739270510] [number_counter]: Counter:  2
[INFO] [1711528816.739186942] [number_counter]: Counter:  4
[INFO] [1711528817.739050485] [number_counter]: Counter:  6
[INFO] [1711528818.738992607] [number_counter]: Counter:  8
```

As you can see, the `number_counter` node adds 2 to the counter every `1.0` second. If you see this, then the publish/subscribe communication between your two nodes is working.

You can start and stop the `number_publisher` node and see that every time you start it, `number_counter` continues to add numbers from the current count.

Writing a C++ subscriber

Let's create the `number_counter` node in C++. The principle is the same, so let's just focus on the syntax here.

Creating a C++ node with a subscriber

Create a new file for your node:

```
$ cd ~/ros2_ws/src/my_cpp_pkg/src/
$ touch number_counter.cpp
```

Open this file and write the code for the node (once again, the complete code is on GitHub).

To create a subscriber in your node, run the following code:

```
number_subscriber_ = this->create_subscription<example_
interfaces::msg::Int64>(
        "number",
        10,
        std::bind(&NumberCounterNode::callbackNumber, this, _1));
```

We find the same components as for Python (but in a different order): topic interface, topic name, queue size, and callback for received messages. For `_1` to work, don't forget to add `using namespace std::placeholders;` before it.

> **Note**
> Even if the `rclpy` and `rclcpp` libraries are supposed to be based on the same underlying code, there can still be some differences in the API. Don't worry if the code sometimes doesn't look the same between Python and C++.

The subscriber object is declared as a private attribute:

```
rclcpp::Subscription<example_interfaces::msg::Int64>::SharedPtr
number_subscriber_;
```

We use the `rclcpp::Subscription` class here, and once again, we create a shared pointer to that object.

We then have the callback method, `callbackNumber`:

```
void callbackNumber(const example_interfaces::msg::Int64::SharedPtr
msg)
{
    counter_ += msg->data;
    RCLCPP_INFO(this->get_logger(), "Counter: %d", counter_);
}
```

The message we receive in the callback is also a (`const`) shared pointer. Hence, don't forget to use `->` when accessing the `data` field.

In this callback, we add the received number to the counter and print it.

Running the C++ subscriber

Create a new executable for that node. Open `CMakeLists.txt` and add the following code:

```
add_executable(number_counter src/number_counter.cpp)
ament_target_dependencies(number_counter rclcpp example_interfaces)

install(TARGETS
  test_node
  number_publisher
  number_counter
  DESTINATION lib/${PROJECT_NAME}/
)
```

Then, build my_cpp_pkg, source the workspace, and run both the publisher and the subscriber node in different Terminals. You should see a similar output to what we had with Python.

Running the Python and C++ nodes together

We've just created a publisher and subscriber for both Python and C++. The topic we use has the same name (`number`) and interface (`example_interfaces/msg/Int64`).

If the topic is the same, it means that you could start the Python `number_publisher` node with the C++ `number_counter` node, for example.

Let's verify that:

```
$ ros2 run my_py_pkg number_publisher
[INFO] [1711597703.615546913] [number_publisher]: Number publisher has
been started.
```

```
$ ros2 run my_cpp_pkg number_counter
[INFO] [1711597740.879160448] [number_counter]: Number Counter has
been started.
[INFO] [1711597741.607444197] [number_counter]: Counter: 2
[INFO] [1711597742.607408224] [number_counter]: Counter: 4
```

You can also try the opposite by running the C++ `number_publisher` node with the Python `number_counter` node.

Why is it working? Simply because ROS 2 is language-agnostic. You could have any node written in any (supported) programming language, and this node could communicate with all the other nodes in the network, using topics and other ROS 2 communications.

ROS 2 communications happen at a lower level, using **Data Distribution Service** (**DDS**). This is the middleware part and is responsible for sending and receiving messages between nodes. When you write a Python or C++ node, you are using the same DDS functionality, with an API implemented in either `rclpy` or `rclcpp`.

I will not go too far with this explanation as it's quite advanced and not really in the scope of this book. If there is just one thing to remember from this, it's that Python and C++ nodes can communicate with each other using ROS 2 communication features. You can create some nodes in Python and other nodes in C++; just make sure to use the same communication name and interface on both sides.

Additional tools to handle topics

You've just written a bunch of nodes containing publishers and subscribers. We will now explore how ROS 2 tools can help you do more things with topics.

We will explore the following topics:

- Introspection with `rqt_graph`
- Introspection and debugging with the `ros2 topic` command line

- Changing a topic name when starting a node
- Replaying topic data with bags

Introspecting topics with rqt_graph

We used `rqt_graph` to visualize nodes in *Chapter 3*. Let's run it again and see how to introspect the publisher and subscriber we have just created.

First, start both the `number_publisher` and `number_counter` nodes (from any package: `my_py_pkg` or `my_cpp_pkg`).

Then, start `rqt_graph` in another Terminal:

```
$ rqt_graph
```

If needed, refresh the view a few times and select **Nodes/Topics (all)**. You can also uncheck the **Dead sinks** box and the **Leaf topics** box. This will allow you to see topics even if there is just one subscriber and no publisher, or one publisher and no subscriber:

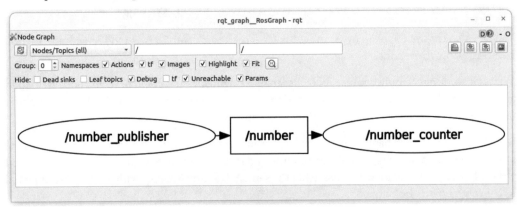

Figure 5.7 – The number topic on rqt_graph

There, we can see the `number_publisher` node and the `number_counter` node. In the middle, we have the `/number` topic, and we can see which node is a publisher or a subscriber.

The `rqt_graph` package can be extremely useful when debugging topics. Imagine that you run some nodes and you're wondering why topic messages are not received by a subscriber. Maybe those nodes are not using the same topic name. You can easily see this with `rqt_graph`:

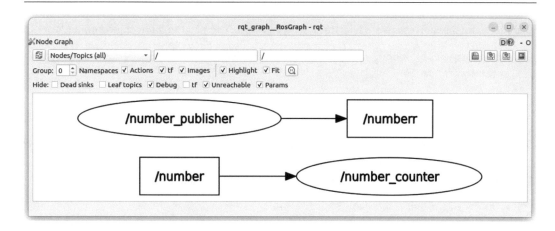

Figure 5.8 – Topic name mismatch between publisher and subscriber

In this example, I made an intentional error in the topic name inside the publisher. Instead of `number`, I have written `numberr`. With `rqt_graph`, I can see where the issue is. The two nodes are not communicating with each other.

The ros2 topic command line

With `ros2 node`, we get additional command-line tools for nodes. For topics, we will use `ros2 topic`.

If you run `ros2 topic -h`, you'll see that there are quite a lot of commands. You already know some of them. Here, I will do a quick recap and explore a few more commands that can be useful when debugging topics.

First, to list all topics, use `ros2 topic list`:

```
$ ros2 topic list
/number
/parameter_events
/rosout
```

As you can see, we get the `/number` topic. You will also always get `/parameter_events` and `/rosout` (all ROS 2 logs are published on this topic).

With `ros2 topic info <topic_name>`, you can get the interface for the topic, as well as the number of publishers and subscribers for that topic:

```
$ ros2 topic info /number
Type: example_interfaces/msg/Int64
Publisher count: 1
Subscription count: 1
```

Then, to go further and see the details for the interface, you can run the following command:

```
$ ros2 interface show example_interfaces/msg/Int64
# some comments
int64 data
```

With this, we have all the information we need to create an additional publisher or subscriber to the topic.

On top of that, we can also directly subscribe to the topic from the Terminal with `ros2 topic echo <topic_name>`. That's what we did just after writing the publisher so that we can make sure it's working before we write any subscriber:

```
$ ros2 topic echo /number
data: 2
---
data: 2
---
```

On the other hand, you can publish to a topic directly from the Terminal with `ros2 topic pub -r <frequency> <topic_name> <interface> <message_in_json>`. To test this, stop all nodes, and start only the `number_counter` node in one Terminal. Apart from the first log, nothing will be printed. Then, run the following command in another Terminal:

```
$ ros2 topic pub -r 2.0 /number example_interfaces/msg/Int64 \
"{data: 7}"
publisher: beginning loop
publishing #1: example_interfaces.msg.Int64(data=7)
publishing #2: example_interfaces.msg.Int64(data=7)
```

This will publish on the `/number` topic at `2.0` Hertz (every `0.5` seconds). When you run this, you'll see some logs on the `number_counter` node, meaning that the messages have been received:

```
[INFO] [1711600360.459298369] [number_counter]: Counter: 7
[INFO] [1711600360.960216275] [number_counter]: Counter: 14
[INFO] [1711600361.459896877] [number_counter]: Counter: 21
```

This way, you can test a subscriber without having to write a publisher first. Note that this only really works for topics with a simple interface. When the interface contains too many fields, it becomes too complicated to write everything on the Terminal.

> **Note**
>
> Both `ros2 topic echo` and `ros2 topic pub` can save you lots of time, and it's also great for collaborating with other people on a project. You could be responsible for writing a publisher, and someone else would write a subscriber. With those command-line tools, both of you can make sure the topic communication is working. Then, when you run the two nodes together, you know that the data you send or receive is correct.

Changing a topic name at runtime

In *Chapter 4*, you learned how to change a node name at runtime—that is, by adding `--ros-args -r __node:=<new_name>` after the `ros2 run` command.

So, for any additional argument you pass after `ros2 run`, add `--ros-args`, but only once.

Then, you can also change a topic name at runtime. To do that, add another `-r`, followed by `<topic_name>:=<new_topic_name>`.

For example, let's rename our topic from `number` to `my_number`:

```
$ ros2 run my_py_pkg number_publisher --ros-args -r number:=my_number
```

Now, if we start the `number_counter` node, to be able to receive the messages, we also need to modify the topic name:

```
$ ros2 run my_py_pkg number_counter --ros-args -r number:=my_number
```

With this, the communication will work, but this time using the `my_number` topic.

To make things a bit more interesting, let's keep those two nodes running, and let's run another publisher to this topic, using the same `number_publisher` node. As you know, we can't have two nodes running with the same name. Thus, we will have to rename both the node and the topic. In a third Terminal, run the following code:

```
$ ros2 run my_py_pkg number_publisher --ros-args -r \
  __node:=number_publisher_2 -r number:=my_number
```

After you run this, you'll see that the `number_counter` receives messages twice as fast since there are two nodes publishing one message every `1.0` second.

On top of that, let's start `rqt_graph`:

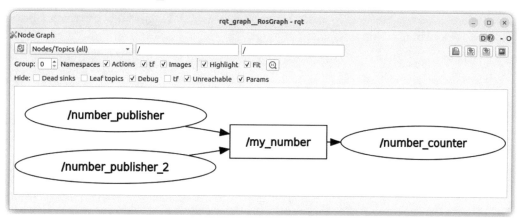

Figure 5.9 – Two publishers and a subscriber, with a renamed topic

We'll see that we have two nodes containing a publisher on the `my_number` topic, and one node containing a subscriber.

Changing topic names at runtime will be quite useful for you, especially when you want to run several existing nodes that you can't modify. Even if you can't rewrite the code, you can modify the names at runtime. Now, let's continue with the tools and explore ROS 2 bags.

Replaying topic data with bags

Imagine this scenario: you're working on a mobile robot that's supposed to perform in a certain way when navigating outside and while it's raining.

Now, this means you will need to run the robot in those conditions so that you can develop your application. There are a few problems: maybe you won't have access to the robot every time, or you can't take it outside, or it's simply not raining every day.

A solution to this is to use ROS 2 bags. Bags allow you to record a topic and replay it later. Thus, you can run the experiment once with the required conditions, and then replay the data just like it was recorded. With this data, you can develop your application.

Let's consider another scenario: you work with a piece of hardware that isn't stable yet. Most of the time, it doesn't work properly. You could record a bag while the hardware is working fine, and then replay this bag to develop your application instead of running the hardware again and again and wasting time on this.

To work with ROS 2 bags, you must use the `ros2 bag` command-line tool. Let's learn how to save and replay a topic with bags.

First, stop all nodes and run the `number_publisher` node only.

We already know that the topic name is `/number`. You can retrieve that with `ros2 topic list` if needed. Then, in another Terminal, record the bag with `ros2 bag record <list of topics> -o <bag_name>`. To make things more organized, I suggest that you create a `bags` folder and record from within this folder:

```
$ mkdir ~/bags
$ cd ~/bags/
$ ros2 bag record /number -o bag1
...
[INFO] [1711602240.190476880] [rosbag2_recorder]: Subscribed to topic '/number'
[INFO] [1711602240.190542569] [rosbag2_recorder]: Recording...
[INFO] [1711602240.190729185] [rosbag2_recorder]: All requested topics are subscribed. Stopping discovery...
```

At this point, the bag is recording and saving all incoming messages inside a database. Let it run for a few seconds, then stop it with *Ctrl + C*:

```
[INFO] [1711602269.786924027] [rosbag2_cpp]: Writing remaining
messages from cache to the bag. It may take a while
[INFO] [1711602269.787416646] [rosbag2_recorder]: Event publisher
thread: Exiting
[INFO] [1711602269.787547010] [rosbag2_recorder]: Recording stopped
```

The `ros2 bag` command will exit, and you'll end up with a new directory named `bag1`. In this directory, you will find a `.mcap` file containing the recorded messages and a YAML file with more information. If you open this YAML file, you'll see the recorded duration, number of recorded messages, and topics that were recorded.

Now, you can replay the bag, which means you'll publish on the topic exactly like it was done when recording.

Stop the `number_publisher` node and replay the bag with `ros2 bag play <path_to_bag>`:

```
$ ros2 bag play ~/bags/bag1/
```

This will publish all the recorded messages, with the same duration as the recording. So, if you record for 3 minutes and 14 seconds, the bag will replay the topic for 3 minutes and 14 seconds. Then, the bag will exit, and you can play it again if you want.

While the bag is playing, you can run your subscriber(s). You can do a quick test with `ros2 topic echo /number` and see the data. You can also run your `number_counter` node, and you will see that the messages are received.

You are now able to save and replay a topic using ROS 2 bags. You can explore more advanced options with `ros2 bag -h`.

As you've seen, there are quite a few available tools for topics. Use these tools as often as possible to introspect, debug, and test your topics. They will save you lots of time when you're developing your ROS 2 application.

We're almost done with topics. So far, all we've done is use existing interfaces. Now, let's learn how to create a custom interface.

Creating a custom interface for a topic

When creating a publisher or subscriber for a topic, you know that you have to use a name and an interface.

It's quite easy to publish or subscribe to an existing topic: you'll find the name and interface using the `ros2` command line, and use that in your code.

Now, if you want to start a publisher or subscriber for a new topic, you will need to choose a name and interface by yourself:

- **Name**: No problem—it's just a chain of characters
- **Interface**: You have two choices—using an existing interface that works with your topic or creating a new one

Let's try to apply the ROS 2 philosophy of not reinventing the wheel. When you create a new topic, check if there is any existing interface that can match your needs. If so, then use it; don't recreate it.

First, you'll learn where you can find existing interfaces. Then, you'll learn how to create a new one.

> **Note**
>
> It's quite common to use the word *message* when talking about topic interfaces. I could have named this section *Creating a custom message*. In the following section, when I talk about messages, I'm referring to topic interfaces.

Using existing interfaces

Before you start a new publisher or subscriber for a topic, take some time to think about what kind of data you want to send or receive. Then, check if an already existing interface contains what you need.

Where to find interfaces

Just like nodes, interfaces are organized in packages. You can find the most common packages for ROS 2 interfaces here: `https://github.com/ros2/common_interfaces`. Not all existing interfaces are listed here, but it's already quite a lot. For other interfaces, a simple search on the internet should bring you to the corresponding GitHub repository.

In this common interfaces repository, you can find the `Twist` message we used with Turtlesim, inside the `geometry_msgs` package. As you can see, for topic interfaces, we then have an additional `msg` folder, which contains all the message definitions for that package.

Now, let's say you want to create a driver node for a camera and publish the images to a topic. If you look inside the `sensor_msgs` package, and then inside the `msg` folder, you'll find a file named `Image.msg`. This *Image* message is probably suitable for your needs. It is also used by a lot of other people, so it will even make your life easier.

Using an existing interface in your code

To use this message, make sure you've installed the package that contains the message—in this case, sensor_msgs. As a quick reminder, to install a ROS 2 package, you can run `sudo apt install ros-<distro>-<package-name>`:

```
$ sudo apt install ros-jazzy-sensor-msgs
```

Maybe the package was already installed. If not, source your environment again afterward. Then, you can find the details regarding the interface with `ros2 interface show <interface>`:

```
$ ros2 interface show sensor_msgs/msg/Image
```

To use this message in your code, just follow what we did in this chapter (with the `example_interfaces/msg/Int64` message):

1. In the `package.xml` file of the package where you write your nodes, add the dependency to the interface package.
2. In your code, import the message and use it in your publisher or subscriber.
3. For C++ only: Add the dependency to the interface package in the `CMakeLists.txt` file.

We will see another example of this process very soon, just after we create our interface.

At this point, you know how to find and use existing messages in your code. But should you always do that?

When not to use existing messages

For common use cases, sensors, and actuators, you will probably find what you need. However, if the interface doesn't match exactly what you want, you will have to create a new one.

There are a few packages containing basic interfaces, such as `example_interfaces`, or even `std_msgs`. You could be tempted to use them in your code. As a best practice, it's better to avoid it. Just read the comments from the message definitions to be sure of that:

```
$ ros2 interface show example_interfaces/msg/Int64
# This is an example message of using a primitive datatype, int64.
# If you want to test with this that's fine, but if you are deploying
it into a system you should create a semantically meaningful message
type.
# If you want to embed it in another message, use the primitive data
type instead.
int64 data
$ ros2 interface show std_msgs/msg/Int64
# This was originally provided as an example message.
# It is deprecated as of Foxy
# It is recommended to create your own semantically meaningful
```

```
message.
# However if you would like to continue using this please use the
equivalent in example_msgs.
int64 data
```

As you can see, the `std_msgs` package is deprecated, and `example_interfaces` is only recommended to make tests—which is what we've done in this chapter so far to help us learn various topics.

As a general rule, if you don't find exactly what you need in the existing interface packages, then create your own interface. It's not hard to do and will always be the same process.

Creating a new topic interface

You will now create your first custom interface for a topic. We will see how to set a package up for that, how to create and build the interface, and how to use it in our code.

Creating and setting up an interfaces package

Before we create any topic interface (message), we need to create a new package and set it up for building interfaces. As a best practice, in your application, you will have one package dedicated to custom interfaces. This means that you create interfaces only in this package, and you keep this package only for interfaces—no nodes or other things, just interfaces. This will make it much easier when you're scaling the application and will help you avoid creating a dependency mess.

A common practice when naming this interface package is to start with the name of your application or robot and add the `_interfaces` suffix. So, if your robot is named `abc`, you should use `abc_interfaces`.

We don't have a robot for this example, so let's just name the package `my_robot_interfaces`.

Create a new package with the `ament_cmake` build type and no dependencies. You don't even need to provide the build type since `ament_cmake` is the one used by default. Navigate to the `src` directory of your workspace and create this package:

```
$ cd ~/ros2_ws/src/
$ ros2 pkg create my_robot_interfaces
```

At this point, your workspace should contain three packages: `my_py_pkg`, `my_cpp_pkg`, and `my_robot_interfaces`.

We need to set this new package up and modify a few things so it can build messages. Go into the package, remove the `src` and `include` directories, and create a new `msg` folder:

```
$ cd my_robot_interfaces/
$ rm -r src/ include/
$ mkdir msg
```

Now, open the `package.xml` file for this package. After `<buildtool_depend>ament_cmake</buildtool_depend>`, add the following three lines. I recommend that you just copy and paste them so that you don't make any mistakes:

```
<build_depend>rosidl_default_generators</build_depend>
<exec_depend>rosidl_default_runtime</exec_depend>
<member_of_group>rosidl_interface_packages</member_of_group>
```

With that, the `package.xml` file is complete and you won't have to do anything else with it for now. Open the `CMakeLists.txt` file. After `find_package(ament_cmake REQUIRED)`, and before `ament_package()`, add the following lines (you can also remove the `if(BUILD_TESTING)` block):

```
find_package(rosidl_default_generators REQUIRED)
rosidl_generate_interfaces(${PROJECT_NAME}
  # we will add the name of our custom interfaces here
)
ament_export_dependencies(rosidl_default_runtime)
```

There's not much to understand about these lines you're adding. They will find some dependencies (`rosidl` packages) and prepare your package so that it can build interfaces.

At this point, your package is ready and you can add new interfaces. You will only need to do this setup phase once. At this point, adding a new interface is very quick.

Creating and building a new topic interface

Let's say we want to create a publisher to send some kind of hardware status for our robot, including the robot version, internal temperature, a flag to know if the motors are ready, and a debug message.

We've looked at existing interfaces and nothing matches. How can you name this new interface? Here are the rules you have to follow:

- Use UpperCamelCase—for example, HardwareStatus
- Don't write `Msg` or `Interface` in the name as this would add unnecessary redundancy
- Use `.msg` for the file extension

Following these rules, create a new file named `HardwareStatus.msg` in the `msg` folder:

```
$ cd ~/ros2_ws/src/my_robot_interfaces/msg/
$ touch HardwareStatus.msg
```

Inside this file, we can add the definition for the message. Here's what you can use:

- Built-in types, such as `bool`, `byte`, `int64`, `float64`, and `string`, as well as arrays of those types. You can find the complete list here: https://docs.ros.org/en/rolling/Concepts/Basic/About-Interfaces.html#field-types.
- Other existing messages, using the name of the package, followed by the name of the message —for example, `geometry_msgs/Twist` (don't add the `msg` folder here).

To make things simple here, we will start with only built-in types. Write the following inside the message file:

```
int64 version
float64 temperature
bool are_motors_ready
string debug_message
```

For each field, we provide the data type, and then the name of the field.

Now, how are we going to build this message? How can we get a Python or C++ class that we can include and use in our code?

To build the message, you simply have to add one line to `CMakelists.txt`, specifying the relative path to the message file:

```
rosidl_generate_interfaces(${PROJECT_NAME}
   "msg/HardwareStatus.msg"
)
```

For each new interface you build in this package, you will add one line inside the `rosidl_generate_interfaces()` function. *Don't add any commas between the lines.*

Now, save all the files and build your new package:

```
$ cd ~/ros2_ws/
$ colcon build --packages-select my_robot_interfaces
Starting >>> my_robot_interfaces
Finished <<< my_robot_interfaces [4.00s]
Summary: 1 package finished [4.28s]
```

The build system will take the interface definition you've written and use it to generate source code for Python and C++:

Creating a custom interface for a topic

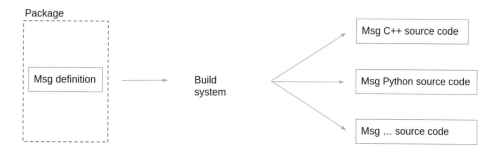

Figure 5.10 – Build system for interfaces

Once you've built the package, make sure you source the environment. You should be able to see your interface from the Terminal (don't forget to use auto-completion to build the command faster and be sure you have the correct name):

```
$ source ~/.bashrc
$ ros2 interface show my_robot_interfaces/msg/HardwareStatus
int64 version
float64 temperature
bool are_motors_ready
string debug_message
```

If you see this, it means that the build process succeeded. If you can't see the interface in the Terminal, then you need to go back and check that you did all the steps correctly.

Using your custom message in your code

Let's say you want to use your new interface in the `number_publisher` node you created in this chapter, inside the my_py_pkg package.

First, open the `package.xml` file from the my_py_pkg package and add a dependency to my_robot_interfaces:

```
<depend>rclpy</depend>
<depend>example_interfaces</depend>
<depend>my_robot_interfaces</depend>
```

Then, for Python, do the following:

1. Import the message by by adding the following import line in your code:

    ```
    from my_robot_interfaces.msg import HardwareStatus
    ```

2. Create a publisher and specify the `HardwareStatus` interface.

3. Create a message in your code, like so:

```
msg = HardwareStatus()
msg.temperature = 34.5
```

> **Note**
> If you're using VS Code, the message might not be recognized after you import it. Close VS Code and open it again in a sourced environment. So, make sure the interface has been built correctly, then source the environment, and open VS code.

If you want to use this message in your C++ node from the my_cpp_pkg package, add the dependency to my_robot_interfaces in the package.xml file of my_cpp_package. Then, do the following:

1. Import the message by adding the following include line in your code:

    ```
    #include "my_robot_interfaces/msg/hardware_status.hpp"
    ```

2. Create a publisher and specify the interface with <my_robot_interfaces::msg::HardwareStatus>.

3. Create a message in your code, like so:

    ```
    auto msg = my_robot_interfaces::msg::HardwareStatus();
    msg.temperature = 34.5;
    ```

When using VS code, the C++ include will not be recognized. You need to add a new line to the c_cpp_properties.json file that was auto-generated (inside a .vscode folder) when you started VS Code. You can find this file from VS Code using the explorer on the left. Then, in the includePath array, add the following line:

```
"includePath": [
        "/opt/ros/jazzy/include/**",
        "/home/<user>/ros2_ws/install/my_robot_interfaces/include/**",
        "/usr/include/**"
    ],
```

You can now create and use your custom interface for topics. As you've seen, first, check whether there's any existing interface that matches your needs. If there is, don't reinvent the wheel. If nothing matches perfectly, however, don't hesitate to create your own interface. To do that, you must create a new package dedicated to interfaces. Once you've finished the setup process for this package, you can add as many interfaces as you want.

Before we wrap things up, I will give you an additional challenge so that you can practice the concepts that were covered in this chapter.

Topic challenge – closed-loop control

Here's a challenge for you so that you can continue practicing creating nodes, publishers, and subscribers. We will start a new ROS 2 project and improve it throughout the following chapters, as we discover more concepts.

I encourage you to read the instructions and take the time to complete this challenge before you check the solution. Practicing is the key to effective learning.

I will not provide a full explanation of all the steps, just a few remarks on the important points. You can find the complete solution code on GitHub, for both Python and C++.

Your challenge is to write a controller for the `turtlesim` node. So far, we've just used simple and basic numbers to publish and subscribe to topics. With this, you can practice as if you were working on a real robot.

Challenge

The goal is simple: we want to make the turtle move in a circle. On top of this, we also want to modify the velocity of the turtle, whether it's on the right or left of the screen.

To get the X coordinate of a turtle on the screen, you can subscribe to the `pose` topic for that turtle. Then, finding the middle of the screen is easy: the minimum X value on the left is 0, and the maximum X value on the right is about `11`. We will assume that the X coordinate for the middle of the screen is `5.5`.

You can then send a command velocity by publishing to the `cmd_vel` topic for the turtle. To make the turtle move in a circle, you just have to publish constant values for the linear X and angular Z velocities. Use `1.0` for both velocities if the turtle is on the left ($X < 5.5$), and `2.0` for both if the turtle is on the right.

Follow these steps to get started:

1. Create a new package (let's name it `turtle_controller`). You can decide to create either a Python or C++ package. If you do both, make sure you give each a different name.
2. Inside this package, create a new node named `turtle_controller`.
3. In the node's constructor, add a publisher (command velocity) and a subscriber (pose).
4. This is where it's a bit different from before: instead of creating a timer and publishing from the timer callback, you can publish directly from the subscriber callback. The `turtlesim` node is constantly publishing on the `pose` topic. Publishing a command from the subscriber callback allows you to create some kind of closed-loop control. You can get the current X coordinate and send a different velocity command, depending on where the turtle is.

To test your code, create an executable out of your code. Then, run `turtlesim` in one Terminal and your node in another. You should see the turtle drawing a circle, with a different velocity depending on where the turtle is.

Solution

You can find the complete code (for both Python and C++) and package organization on GitHub.

Here are the most important steps for the Python node. The code starts with all the required import lines:

```python
#!/usr/bin/env python3
import rclpy
from rclpy.node import Node
from geometry_msgs.msg import Twist
from turtlesim.msg import Pose
```

Here, we import `Twist` from `geometry_msgs` and `Pose` from `turtlesim`. You can find those interfaces by running `turtlesim_node` and exploring topics with the `ros2 topic` and `ros2 interface` command-line tools.

Then, we create a class for our node, with a constructor:

```python
class TurtleControllerNode(Node):
    def __init__(self):
        super().__init__("turtle_controller")
        self.cmd_vel_pub_ = self.create_publisher(Twist, "/turtle1/cmd_vel", 10)
        self.pose_sub_ = self.create_subscription(Pose, "/turtle1/pose", self.callback_pose, 10)
```

As you can see, we just create a publisher and a subscriber. There's no timer as we plan to use the publisher directly from the subscriber callback:

```python
def callback_pose(self, pose: Pose):
    cmd = Twist()
    if pose.x < 5.5:
        cmd.linear.x = 1.0
        cmd.angular.z = 1.0
    else:
        cmd.linear.x = 2.0
        cmd.angular.z = 2.0
    self.cmd_vel_pub_.publish(cmd)
```

This is the subscriber callback. Whenever we receive a new `Pose` message, we create a new command (a `Twist` message). Then, depending on the current *X* coordinate of the turtle, we give different values for the velocity. Finally, we publish the new velocity command.

That's it for this challenge. It can be a bit challenging to understand how to start, but in the end, you can see that there is not so much code to write. I encourage you to come back to this challenge in a few days and try again without looking at the solution. This way, you can check if you understood the concept of topics correctly.

Summary

In this chapter, you worked on ROS 2 topics.

Topics allow nodes to communicate with each other using a publish/subscribe mechanism. Topics are made for unidirectional data streams and are anonymous.

You can write topic publishers and subscribers directly in your nodes by using `rclpy` for Python and `rclcpp` for C++.

To write a publisher, you must do the following:

1. First, check what topic name and interface you must send. Import the interface into the code and create a publisher in the node's constructor.
2. To publish, you must create a message, fill in the different fields, and publish the message with your publisher.

You can potentially publish a message from anywhere in the code. A common structure is to add a timer and publish from the timer callback. If it makes sense, you can also publish from a subscriber callback directly.

To write a subscriber, you must do the following:

1. As for the publisher, you need to know what name and interface to receive. Import the interface and create a subscriber in the node's constructor.
2. When creating the subscriber, you will need to specify a callback function. It's in this callback function that you can receive and process incoming messages.

If you create a publisher or subscriber for a new topic and no interface matches your needs, you might need to create a custom interface. In this case, you must do the following:

1. Create and configure a new package dedicated to interfaces for your robot or application.
2. Add your topic interface inside the package and build the package.
3. Now, you can use this custom interface in your publishers/subscribers, just like any other interface.

To try a publisher or a subscriber, simply build the package where the node is, source the environment, and run the node. You can then use the `ros2` command-line tools, as well as `rqt_graph`, to introspect your application and solve potential issues.

After topics, the next logical step is to learn about ROS 2 services. This is what we will cover in the following chapter.

6
Services – Client/Server Interaction between Nodes

Nodes can communicate with each other using one of three communication types. You discovered topics in the previous chapter. Now is the time to switch to the second most used communication: ROS 2 services.

As we did for topics, we will first understand services with the help of a real-life analogy. I will also share more thoughts on when to use topics versus services. After that, you will dive into the code and write a service server and client inside nodes using custom service interfaces. You will also explore additional tools to handle services from the Terminal.

All the code we'll write in this chapter starts from the final code of the previous chapter. We will improve the number application to learn how to use services, and then work on the turtle controller application with an additional challenge. If you want to have the same starting point as me, you can download the code from GitHub (`https://github.com/PacktPublishing/ROS-2-from-Scratch`), in the `ch5` folder, and use it as a starting point. The final code can be found in the `ch6` folder.

By the end of this chapter, you will understand how services work, and you will be able to create your own service interfaces, service servers, and service clients.

Becoming confident with topics and services is one of the most important things when starting with ROS 2. With this, you will be able to write custom code for your projects and interact with most of the existing ROS 2 applications.

In this chapter, we will cover the following topics:

- What is a ROS 2 service?
- Creating a custom service interface
- Writing a service server

- Writing a service client
- Additional tools to handle services
- Service challenge – client and server

What is a ROS 2 service?

You discovered the concept of ROS 2 services in *Chapter 3*, in the *Services* section, where you ran your first service server and client to get an intuition of how they work. You also became familiar with the `ros2` command-line tool for handling services from the Terminal.

From here, I will start from scratch again and explain what services are, using a real-life analogy. We will build an example, step by step, and then recap the most important points.

A server and a client

To start, I will use an online weather service as an analogy.

This online weather service can tell us the local weather after we send our location. To get the weather report for your city, you will need to interact with this service. You can use your computer to send a web request with the URL provided by the service.

What's going to happen? First, your computer will send a request to the weather service. The request contains your location. The service will receive the request, process it, and if the location is valid, it will return the weather for that location. Your computer then receives a response containing the weather information. That's the end of the communication. Here's an illustration of this process:

Figure 6.1 – Client/server interaction

This is basically how a ROS 2 service works. On one side, you have a **service server** inside a node, and on the other side, you have a **service client** inside another node.

To start the communication, the service **Client** needs to send a **request** to the **Server**. The **Server** will then process the request, do any appropriate actions or computations, and return a **response** to the **Client**.

As you can see, a service, just like for topics, has a name and an interface. The interface is not just one message, it's a pair of messages: a request and a response. Both the client and server must use the same name and interface to successfully communicate with each other.

With this example, the HTTP URL is the service name, and the pair (location, weather) is the service interface (request, response).

Multiple clients for one service

In real life, many people will try to get the weather from this online service (at different times or at the same time). That's not a problem: each client will send a request with a location to the server through the HTTP URL. The server will process each request individually and return the appropriate weather information to each client.

Now, this is very important: there can be only one server. One URL only goes to one server, just like one physical mail address is unique. Imagine if you send a package to someone and there are two places with the same address. How can the mail delivery person know where to deliver the package?

This will be the same for ROS 2 services. You can have several clients send a request to the same service. However, for one service, only one server can exist. See the following figure:

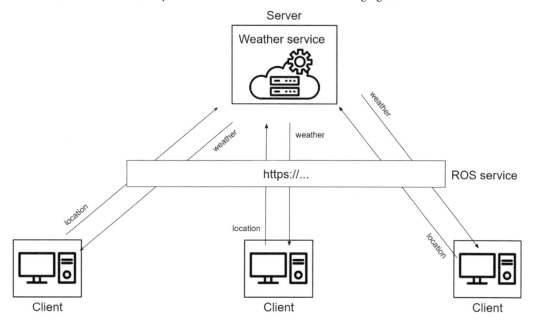

Figure 6.2 – Service server with multiple clients

Here, you can see some boxes, each box representing a node. Thus, we have four nodes. Three nodes contain a service **Client** and talk to the **Weather service** node, which contains a service **Server**.

One thing to note here is that the clients don't know exactly which node to communicate with. They must go through the URL (service name). In this example, the clients aren't aware of the IP address of the server—they just know they have to use the URL to connect to the server.

Also, no client is aware of the other clients. When you try to get the weather information from this service, you don't know who is also trying to access the service, or even how many people are sending a request.

Another service example with robotics

Let's use another example that could be part of a ROS application.

Imagine that you have a node responsible for controlling an LED panel (three LEDs). This node could contain a service server that allows other nodes to request turning an LED on or off.

You also have a node monitoring a battery. In your application, what you want to do is turn on one LED when the battery is low, and then turn it off when the battery is high again.

You can do that using a ROS 2 service. The LED panel node would contain a service server named `set_led`. To send a request to this server, you must provide the LED number and the state of that LED (on or off). Then, you receive a response containing a boolean value to see if the request was successfully processed by the server.

So, the battery is now running low. Here's what's going to happen:

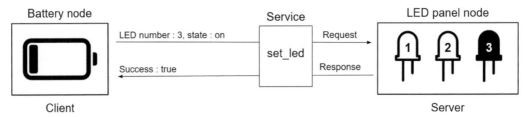

Figure 6.3 – Client asking to turn on LED number 3

Here, **Battery node** will send a **Request** to the `set_led` service. The **Request** contains the **LED number 3** and **state on** details so that it can turn on LED **3** of the panel.

The **Service** server, in the **LED panel node**, receives the **Request**. The server may decide to validate the **Request** (for example, if the LED number is 4, this is not valid) and process it. Processing the **Request** here means turning on the third LED. After that, the server sends a **Response** back to the **Client**, with a boolean flag. The **Client** receives this **Response**, and the communication ends.

Then, when the battery is fully charged, the **Battery node** sends another **Request** this time to turn off **LED 3**:

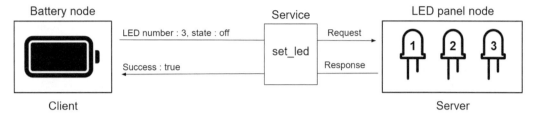

Figure 6.4 – Client asking to turn off LED number 3

The process is the same. The **Client** sends a **Request**, this time with **state off** for **LED 3**. The **Server**, inside the **LED panel node**, receives that **Request** and turns off the third LED. Then, the **Server** sends a **Response** back to the **Client**.

Wrapping things up

On top of topics, ROS 2 nodes can use services to communicate with each other.

When should you use topics versus services? You should use topics to publish unidirectional data streams and services when you want to have a client/server type of communication.

For example, if you want to continuously send a velocity command to a robot 10 times per second, or send the data you read from a sensor, you will use topics. If you want to have a node perform quick computations or do some actions on demand (enabling/disabling a motor, starting/stopping a robot), then you would use services.

It can be hard to give a definitive answer to that question. Each application is different. Most of the time, the choice will be obvious, but sometimes, you have to go one way only to realize that that was the wrong way. The more experience you get with ROS 2, the more you will be able to make the best design decisions.

Here are some important points about how services work:

- A service is defined by a name and an interface.
- The name of a service follows the same rules as for topics. It must start with a letter and can be followed by other letters, numbers, underscores, tildes, and slashes.
- The interface contains two things: a request and a response. Both the client and server must use the same interface to be able to communicate with each other.
- A service server can only exist once but can have multiple clients.

- Service clients are not aware of each other and are not aware of the server node. To reach the server, they just know that they must use the service name and provide the correct interface.
- One node can contain multiple service servers and clients, each with a different service name.

Now, how can you write a service client and server?

Just as for nodes and topics, you will find everything you need in the `rclpy` and `rclcpp` libraries. With those libraries, you can write a service server and client inside nodes. That's what we are going to do now.

As we can't test a client without a server, let's start with the server side. But before we even start writing the server, what interface will we need to use for the service?

Creating a custom service interface

In *Chapter 5*, when we created the ROS 2 application with the `number_publisher` and `number_counter` nodes, we used an existing interface for the `number` topic. Since we wanted to publish an integer number, the `example_interfaces/msg/Int64` interface seemed to be exactly what we needed. At this point, you know that you must avoid using the `example_interfaces` package for real applications, but for a first test, that wasn't a problem.

We're going to continue working on this application and add more functionalities so that we can practice with services. Here, we will focus on the `number_counter` node. For now now, in this node, every time we receive a message from the `number` topic, we'll add this number to a counter and print the counter.

What we want to do is allow the `number_counter` node to reset the counter to a given number when we ask it to. For that, we will add a service server inside the node. Then, any other node can send a request, specifying the reset value for the counter. For example, let's say the counter is currently at 76, and you send a request to reset it to 20. If the request is accepted by the service server, the counter will now become 20 and continue to increment from that value.

Great—we know what we must do. Now, which interface should we use? Can we find an existing interface for what we need, or will we have to create a custom one? As per the title of this section, you can already guess the answer to that question. Nonetheless, let's see what we could find if we were looking at existing interfaces.

Finding an existing interface for our service

When it comes to a service interface, we need to think about two things: the request and the response.

In our application, the request, which is sent from the client to the server, should contain an integer number. This is the reset value for the counter.

For the response, which is sent from the server to the client, we can decide to use a boolean flag, to specify whether we were able to perform the request, and a message to explain what went wrong if something went wrong.

The question is, will we find an existing interface that matches our needs? Unfortunately, this time, it seems that there is no matching interface. We can check the `example_interfaces` package:

```
$ ros2 interface list | grep example_interfaces/srv
example_interfaces/srv/AddTwoInts
example_interfaces/srv/SetBool
example_interfaces/srv/Trigger
```

We can even check the `std_srvs` package:

```
$ ros2 interface list | grep std_srvs/srv
std_srvs/srv/Empty
std_srvs/srv/SetBool
std_srvs/srv/Trigger
```

> **Note**
> As you can see, service interfaces are placed inside a `srv` folder, inside the package. For topics, we had a `msg` folder. This is a good way to differentiate both types of interfaces easily.

If you look more closely at those interfaces, especially `SetBool` and `Trigger`, you'll see that there is no way to send an integer number in the request. Here's an example where we're trying to use `SetBool`:

```
$ ros2 interface show example_interfaces/srv/SetBool
# some comments
bool data # e.g. for hardware enabling / disabling
---
bool success    # indicate successful run of triggered service
string message # informational, e.g. for error messages
```

When looking at the interface definition, you can see that the request and response are separated by three dashes (---). In the response, we can find a boolean and a string, which is what we want. However, the request only contains a boolean, not an integer.

You could have a look at other interfaces in the common interfaces GitHub repository (https://github.com/ros2/common_interfaces) but you won't find exactly what we are looking for.

Thus, we will create our own service interface before writing the code for the service. For the `number` topic, we were lucky enough to find an interface that we could directly use in the code (even though for real applications, the best practice is to avoid using `example_interfaces` and `std_srvs` anyway). Here, we need to create the interface first.

Creating a new service interface

To create a service interface, just like for a topic interface, you need to create and configure a package dedicated to interfaces.

Good news: we did that in *Chapter 5* in the *Creating a custom interface for a topic* section. Since we're working on the same application, we will put all the topic and service interfaces in the same package: `my_robot_interfaces` (if you don't already have this package, go back to the previous chapter and set it up).

We can create a new service interface inside that package directly; there's nothing else to do. So, the process will be quite quick.

First, navigate inside the `my_robot_interfaces` package (where you already have a `msg` folder) and create a new `srv` folder:

```
$ cd ~/ros2_ws/src/my_robot_interfaces/
$ mkdir srv
```

In this new folder, you will put all the service interfaces that are specific to your application (or robot).

Now, create a new file for a service. Here are the rules to follow regarding the filename:

- Use UpperCamelCase (PascalCase)—for example, `ActivateMotor`.
- Don't write `Srv` or `Interface` in the name as this would add unnecessary redundancy.
- Use `.srv` for the file extension.
- As a best practice, use a verb in the interface name—for example, `TriggerSomething`, `ActivateMotor`, or `ComputeDistance`. Services are about doing an action or computation, so by using a verb, you make it very clear what the service is doing.

Since we want to reset the counter, let's simply call the interface `ResetCounter`:

```
$ cd ~/ros2_ws/src/my_robot_interfaces/srv/
$ touch ResetCounter.srv
```

Open this file and write the definition for the service interface. One very important thing to do here is add three dashes (- - -) and put the request definition on top, and then the response definition below the dashes.

For the request and response, you can use the following:

- Built-in types (`bool`, `byte`, `int64`, and so on).
- Existing message interfaces. For example, the request of the service could contain `geometry_msgs/Twist`.

> **Note**
>
> You can't include a service definition inside another service definition. You can only include a message (topic definition) inside the request or the response of the service. The request and response can be seen as two independent messages.

Let's write our service interface. As it's not too complex, we can use simple built-in types:

```
int64 reset_value
---
bool success
string message
```

With this, the client will send a request with one integer value, and the response will contain a boolean flag as well as a string. All the fields inside the definition must follow the snake_case convention (use underscores between words, all letters lowercase, and no space).

> **Note**
>
> Make sure you always have three dashes in all your service definitions, even if the request or the response is empty.

Now that we've written our interface, we need to build it so that we can use it in our code.

Go back to the `CMakeLists.txt` of `my_robot_interfaces` package. Since the package has already been configured, we just need to add one line. Add the relative path to the interface on a new line inside the `rosidl_generate_interfaces()` function. Don't use any commas between the lines:

```
rosidl_generate_interfaces(${PROJECT_NAME}
  "msg/HardwareStatus.msg"
  "srv/ResetCounter.srv"
)
```

After this, save all files and build the `my_robot_interfaces` package:

```
$ colcon build --packages-select my_robot_interfaces
```

Once built, source the environment. You should be able to find your new interface:

```
$ ros2 interface show my_robot_interfaces/srv/ResetCounter
int64 reset_value
---
bool success
string message
```

If you see that, you know that the service interface has been built successfully, and you can use it in your code. So, let's do that and write a service server.

Writing a service server

You will now write your first service server. As mentioned previously, we will continue with the number application we started in the previous chapter. What we want to do here is allow `number_counter` to reset the counter to a given number when we ask it to do so. This is a perfect example of when to use a service.

The first thing to think about when creating a new service is what service interface you need. We've just done that, so we can now focus on the code.

To write a service server, you will need to import the interface and then create a new service in the node's constructor. You will also need to add a callback to be able to process the request, do the required action or computation, and return a response to the client.

As always, let's start with a fully detailed explanation with Python, after which we will see how to do the same with C++.

Writing a Python service server

To write a Python service server, we first need to have a Python node. We won't create a new node here since we're adding functionality to an existing one (`number_counter`).

> **Note**
> You can have any number of publishers, subscribers, and services inside a node. So long as you keep things clean, that will not be a problem.

Let's get started. As always, you can find the complete code in this book's GitHub repository. I will not display the full code for the node here, just the new lines that are required for the service we are adding.

Importing a service interface

The first big part of creating a service is to find an existing interface or create a new one. That's what we just did, so let's use the `ResetCounter` interface from the `my_robot_interfaces` package.

First, we need to add the dependency to this interface package inside the package where we write the node with the service. Open the `package.xml` file from `my_py_pkg` and add the new dependency:

```
<depend>rclpy</depend>
<depend>example_interfaces</depend>
<depend>my_robot_interfaces</depend>
```

This will ensure that the interfaces package is installed when you build the `my_py_pkg` package with `colcon`. Now, import the dependency into your code (`number_counter.py`):

```
#!/usr/bin/env python3
import rclpy
from rclpy.node import Node
from example_interfaces.msg import Int64
from my_robot_interfaces.srv import ResetCounter
```

To import a service, we must specify the package name (`my_robot_interfaces`), followed by the folder name for services (`srv`), and finally the class for the interface (`ResetCounter`).

> **Note**
>
> I've already mentioned this, but if you're using VS Code and auto-completion doesn't work, or the service isn't recognized (import error), follow the process below.
>
> Close VS code. Then, open a new Terminal, make sure the environment is correctly sourced, and find the interface (`ros2 interface show <interface_name>`). After, navigate to the `src` directory of the ROS 2 workspace and open VS Code with:
>
> `$ code .`

Adding a service server to the node

Now that you've correctly imported the service interface, you can create the service server.

As you did for publishers and subscribers, you will add your service servers to the node's constructor.

Here's the constructor of the `NumberCounterNode` class, which contains the previously created subscriber and the new service server:

```
def __init__(self):
    super().__init__("number_counter")
    self.counter_ = 0
    self.number_subscriber_ = self.create_subscription(Int64,
"number", self.callback_number, 10)
    self.reset_counter_service_ = self.create_service(ResetCounter,
"reset_counter", self.callback_reset_counter)
    self.get_logger().info("Number Counter has been started.")
```

We add the service server at the same time as the number subscriber and just before the ending log.

To create the service server, we use the `create_service()` method from the `Node` class. Once again, you can see that by inheriting from this class, we get access to all ROS 2 functionalities easily. In this method, you must provide three arguments:

- **Service interface**: This is the `ResetCounter` class we have imported.
- **Service name**: Whenever you create a service server, you're creating the service itself, so you decide what its name will be. As best practice, start with a verb. Since we want to reset the counter, we'll simply name it `reset_counter`.
- **Service callback**: The service server, as its name suggests, is a server. This means that it won't do anything by itself. You will need to have a client send a request so that the server does something. So, while the node is spinning, the server will be in "waiting mode." Upon reception of a request, the service callback will be triggered, and the request will be passed to this callback.

Now, we need to implement this callback. First, let's write a minimal code example:

```
def callback_reset_counter(self, request: ResetCounter.Request,
response: ResetCounter.Response):
    self.counter_ = request.reset_value
    self.get_logger().info("Reset counter to " + str(self.counter_))
    response.success = True
    response.message = "Success"
    return response
```

In a service callback, we receive two things: an object for the request and an object for the response. The request object contains all the data sent by the client. The response object is empty, and we will need to fill it, as well as return it.

To name the callback, I usually write `callback_` followed by the service name. This makes it easier to recognize in the code and will prevent future mistakes as you want to make sure you don't call this method directly. It should only be called while the node is spinning and when a client sends a request from another node.

> **Note**
> In the method's arguments, I have also specified the type for the two arguments. This way, we make the code more robust, and we can use auto-completion features from IDEs such as VS Code.
>
> When you create an interface for a topic, you only get one class for that interface (for example, `Int64`). As you can see, in a service, we get two classes: one for the request (`Interface.Request`) and one for the response (`Interface.Response`).

In this callback, we get `reset_value` from the request and modify the `counter_` variable accordingly. Then, we fill the success and message fields from the response and return the response.

This is a very minimal piece of code for a service server. In real life, you'll probably want to check if the request is valid before you use the values from it. For example, if you have a service that will modify the maximum velocity of a mobile robot, you might want to be sure the value you receive is not too high, to prevent the robot from becoming uncontrolled and damaging itself or the environment.

Let's improve the callback so that we can validate `reset_value` before we modify the `counter_` variable.

Validating the request

Let's say we want to add those two validation rules: the reset value must be a positive number, and it cannot be higher than the current counter value.

Modify the code in the `callback_reset_counter` method, like so:

```python
def callback_reset_counter(self, request: ResetCounter.Request,
response: ResetCounter.Response):
    if request.reset_value < 0:
        response.success = False
        response.message = "Cannot reset counter to a negative value"
    elif request.reset_value > self.counter_:
        response.success = False
        response.message = "Reset value must be lower than current counter value"
    else:
        self.counter_ = request.reset_value
        self.get_logger().info("Reset counter to " + str(self.counter_))
        response.success = True
        response.message = "Success"
    return response
```

First, we check if the value is negative. If so, we don't do anything with the `counter_` variable. We set the boolean flag to `False` and provide an appropriate error message.

Then, we check if the value is greater than the current `counter_` value. If that's the case, we do the same thing as before, with a different error message.

Finally, if none of those conditions are true (which means we've validated the request), then we process the request and modify the `counter_` variable.

Here's a recap of the steps for a service server callback:

1. (Optional but recommended) Validate the request, or validate that external conditions are met for the callback to be processed. For example, if the service is about activating a motor, but the communication with the motor hasn't been started yet, then you can't activate the motor.
2. Process the action or computation using the data from the request if needed.

3. Fill in the appropriate field for the response. It's not mandatory to fill in all the fields. If you omit some of them, default values will be used (0 for numbers and " " for strings).
4. Return the response. This is quite an important step that many people forget at the beginning. If you don't return the response, you will get an error at runtime.

All you must do now is build your package where the node is, source, and run the node.

When you run the `number_counter` node, you'll see the following:

```
$ ros2 run my_py_pkg number_counter
[INFO] [1712647809.789229368] [number_counter]: Number Counter has been started.
```

The service server has been started within the node, but of course, nothing will happen as you need to send a request from a client to try the server.

That's what we will do in a minute, but before that, let's learn how to write the service server in C++. If you don't want to learn ROS 2 with C++ for now, you can skip this and go to the next section in this chapter.

Writing a C++ service server

Let's add a service server inside our C++ `number_counter` node using the same name and interface that we used for the one we created with Python. The process is the same: import the interface, create a service server, and add a callback function.

As mentioned previously in this book, make sure you follow all C++ explanations while keeping the GitHub code open on the side.

Importing a service interface

First, since we'll have a dependency on `my_robot_interfaces`, open the `package.xml` file of the `my_cpp_pkg` package and add the following one line:

```
<depend>rclcpp</depend>
<depend>example_interfaces</depend>
<depend>my_robot_interfaces</depend>
```

Then, open the `number_counter.cpp` file and include the `ResetCounter` interface:

```
#include "rclcpp/rclcpp.hpp"
#include "example_interfaces/msg/int64.hpp"
#include "my_robot_interfaces/srv/reset_counter.hpp"
```

To import a service interface in C++, you must use `#include "<package_name>/srv/<service_name>.hpp"`.

> **Note**
>
> As a reminder, for this include to be recognized by VS Code, make sure you add the following to the `c_cpp_properties.json` file, in the `.vscode` folder that was generated when you opened VS Code: `"/home/<user>/ros2_ws/install/my_robot_interfaces/include/**"`.

After this, I added an extra line with the `using` keyword so that we can just write `ResetCounter` in the code, instead of `my_robot_interfaces::srv::ResetCounter`:

```
using ResetCounter = my_robot_interfaces::srv::ResetCounter;
```

This will help us make the code more readable. With C++, you can quickly end up with very long types that almost take more than one line to write. Since we will need to use the service interface quite often, adding this `using` line is a best practice to keep things simple.

I didn't do it previously with `example_interfaces::msg::Int64` when we worked on topics, but if you want, you can also write `using Int64 = example_interfaces::msg::Int64;` and then reduce the code for the subscriber.

Adding a service server to the node

Now that we've included the interface, let's create the service server. We will store it as a private attribute in the class:

```
rclcpp::Service<ResetCounter>::SharedPtr reset_counter_service_;
```

As you can see, we use the `rclcpp::Service` class, and then, as always, we make it a shared pointer with `::SharedPtr`.

Now, we can initialize the service in the constructor:

```
reset_counter_service_ = this->create_service<ResetCounter>("reset_
counter", std::bind(&NumberCounterNode::callbackResetCounter, this,
_1, _2));
```

To create the service, we must use the `create_service()` method from the `rclcpp::Node` class. As for Python, we need to provide the service interface, the service name, and a callback to process the incoming requests. For `_1` and `_2` to work, don't forget to add `using namespace std::placeholders;` beforehand.

Here's the callback method, including the code to validate the request:

```
void callbackResetCounter(const ResetCounter::Request::SharedPtr
request, const ResetCounter::Response::SharedPtr response)
{
    if (request->reset_value < 0) {
        response->success = false;
```

```
            response->message = "Cannot reset counter to a negative 
value";
        }
        else if (request->reset_value > counter_) {
            response->success = false;
            response->message = "Reset value must be lower than current 
counter value";
        }
        else {
            counter_ = request->reset_value;
            RCLCPP_INFO(this->get_logger(), "Reset counter to %d", 
counter_);
            response->success = true;
            response->message = "Success";
        }
    }
```

In the callback, we receive two arguments—the request and the response. Both are `const` shared pointers.

What we do in this callback is the same as for Python. The biggest difference here is that we don't return anything (in Python, we had to return the response) as the return type for the callback is `void`.

Now, we can build the package to compile and install the node. However, before we run `colcon build`, we have to modify the `CMakeLists.txt` file of the `my_cpp_pkg` package. Since we have a new dependency on `my_robot_interfaces`, we need to link the `number_counter` executable with that dependency.

First, add a line under all the `find_package()` lines:

```
find_package(ament_cmake REQUIRED)
find_package(rclcpp REQUIRED)
find_package(example_interfaces REQUIRED)
find_package(my_robot_interfaces REQUIRED)
```

Then, add `my_robot_interfaces` to the `ament_target_dependencies()` function, for the `number_counter` executable:

```
add_executable(number_counter src/number_counter.cpp)
ament_target_dependencies(number_counter rclcpp example_interfaces 
my_robot_interfaces)
```

For every new dependency you're using in this executable, you will have to link to it before you build.

If you forget about this, then you will get this kind of error when you run `colcon build`:

```
fatal error: my_robot_interfaces/srv/reset_counter.hpp: No such file 
or directory
Failed   <<< my_cpp_pkg [1.49s, exited with code 2]
```

Now you can build the C++ package, source, and run the `number_counter` node.

```
$ ros2 run my_cpp_pkg number_counter
[INFO] [1712726520.316615636] [number_counter]: Number Counter has been started.
```

We are now at the same point as when we finished the Python service server. The next step is to try the service server. To do that, we need a service client.

Writing a service client

For service communication to work, you need a service server and a service client. As a reminder, you can only have one service server but multiple clients.

So far, we've finished our service server inside the `number_counter` node. Now, let's create a service client inside another node so that you can try the service.

Where will you write the code for the client? In a real application, you will create a service client in a node that needs to call the service. In terms of the battery and LED example from the beginning of this chapter, the LED panel node contains a service server. The battery node, which is responsible for monitoring the battery state, contains a service client that can send some requests to the server.

Then, when to send a request depends on the application. With the previous example, we decided that when the battery gets full or empty, we use the service client inside the node to send a request to the server so that we can turn an LED on/off.

To keep things simple for now, we will create a new node named `reset_counter_client`. This node will only do one thing: send a request to the service server and get the response. With this, we will be able to focus only on writing the service client. As usual, we'll start with Python and then see the C++ code as well.

Writing a Python service client

Create a new file, named `reset_counter_client.py`, inside the `my_py_pkg` package. Make this file executable. The file should be placed with all the other Python files you created previously.

Open the file and start by importing the interface:

```
from my_robot_interfaces.srv import ResetCounter
```

In the node's constructor, create a service client:

```
def __init__(self):
    super().__init__("reset_counter_client")
    self.client_ = self.create_client(ResetCounter, "reset_counter")
```

To create a service client, we use the `create_client()` method from the `Node` class. We need to provide the service interface and service name. Make sure you use the same name and interface you defined in the server.

Then, to call the service, we create a new method:

```python
def call_reset_counter(self, value):
    while not self.client_.wait_for_service(1.0):
        self.get_logger().warn("Waiting for service...")
    request = ResetCounter.Request()
    request.reset_value = value
    future = self.client_.call_async(request)
    future.add_done_callback(
        self.callback_reset_counter_response)
```

Here are the steps to make a service call:

1. Make sure the service is up and running with `wait_for_service()`. This function will return `True` as soon as the service has been found, or return `False` after the provided timeout, which is `1.0` seconds here.
2. Create a request object from the service interface.
3. Fill in the request fields.
4. Send the request with `call_async()`. This will give you a Python `Future` object.
5. Register a callback for when the node receives the response from the server.

To process the response from the service, add a callback method:

```python
def callback_reset_counter_response(self, future):
    response = future.result()
    self.get_logger().info("Success flag: " + str(response.success))
    self.get_logger().info("Message: " + str(response.message))
```

In the callback, we get the response with `future.result()`, and we can access each field of the response. In this example, we simply print the response with a log.

So, what's going to happen? After you send the request with `call_async()`, the server will receive and process the request. Upon completion of the task, the server will return a response to the node where the client is. When the client node receives the response, it will process it in the callback that you've written.

> **Note**
>
> You might be wondering, why do we need a callback? Why can't we just wait for the response in the same method where we send the request? That's because if you block this method (or in other words, this thread), then the node won't be able to spin. If the spin is blocked, then any response you get for this node won't be processed, and you have what is called a deadlock.

The only thing left to do is call the `call_reset_counter()` method. If we don't call it, nothing will happen. In a real application, you would call this method whenever you need it (it could be from a timer callback, a subscriber callback, and so on). Here, to make a test, we just call the method after creating the node, and before spinning, in the `main()` function:

```
node = ResetCounterClientNode()
node.call_reset_counter(20)
rclpy.spin(node)
```

The service client will send a request and register a callback for the response. After that, the `call_reset_counter()` method exits, and the node starts to spin.

That's it for the code. You can use this structure for the client (one method to send the request and one callback to process the response) in any other node.

Now, let's test the client/server communication.

Running the client and server nodes together

Create an executable in the `setup.py` file named `reset_counter_client`, for example.

Then, build the workspace and open three Terminals. In Terminals 1 and 2, start `number_publisher` and `number_counter`. The latter will start the `reset_counter` service server.

In Terminal 3, start the `reset_counter_client` node. Since we want to reset the counter to 20, if the counter inside the `number_counter` node is less than 20 at the moment of sending the request, you will get the following response:

```
$ ros2 run my_py_pkg reset_counter_client
[INFO] [1713082991.940407360] [reset_counter_client]: Success flag: False
[INFO] [1713082991.940899261] [reset_counter_client]: Message: Reset value must be lower than current counter value
```

If the counter is 20 or more, you will get the following response instead:

```
$ ros2 run my_py_pkg reset_counter_client
[INFO] [1713082968.101789868] [reset_counter_client]: Success flag: True
[INFO] [1713082968.102277613] [reset_counter_client]: Message: Success
```

Also, just after you start the node, the client sometimes needs a bit of time to find the service. In this case, you might see this log as well:

```
[WARN] [1713082991.437932627] [reset_counter_client]: Waiting for
service...
```

On the server side (the `number_counter` node), if the counter is being reset, you will see this log:

```
[INFO] [1713083108.125753986] [number_counter]: Reset counter to 20
```

With that, we have tested two cases: when the counter is less than the requested reset value and when the counter is more than the requested reset value. If you want, you can also test the third case: when the requested reset value is lower than 0.

Now that we've finalized the client/server communication between the two nodes, let's switch to C++.

Writing a C++ service client

The C++ code follows the same logic as the Python code.

First, we include the interface:

```cpp
#include "my_robot_interfaces/srv/reset_counter.hpp"
```

Then, we add a few `using` lines to reduce the code later:

```cpp
using ResetCounter = my_robot_interfaces::srv::ResetCounter;
using namespace std::chrono_literals;
using namespace std::placeholders;
```

Next, we declare the service client as a private attribute in the class:

```cpp
rclcpp::Client<ResetCounter>::SharedPtr client_;
```

After, we initialize the client in the constructor:

```cpp
ResetCounterClientNode() : Node("reset_counter_client")
{
    client_ = this->create_client<ResetCounter>("reset_counter");
}
```

Then, as we did for Python, we add a method to call the service:

```cpp
void callResetCounter(int value)
{
    while (!client_->wait_for_service(1s)) {
        RCLCPP_WARN(this->get_logger(), "Waiting for the server...");
    }
```

```
    auto request = std::make_shared<ResetCounter::Request>();
    request->reset_value = value;
    client_->async_send_request(request,
std::bind(&ResetCounterClientNode::callbackResetCounterResponse, this,
_1));
}
```

In this method, we wait for the service (don't forget the exclamation mark in front of `client->wait_for_service(1s)`), create a request, fill in the request, and send it with `async_send_request()`. We pass the callback method as an argument, which will register the callback when the node is spinning.

Here's the callback method for the response:

```
void callbackResetCounterResponse(
    rclcpp::Client<ResetCounter>::SharedFuture future)
{
    auto response = future.get();
    RCLCPP_INFO(this->get_logger(), "Success flag: %d, Message: %s",
(int)response->success, response->message.c_str());
}
```

Finally, to be able to send a request, we call the `callResetCounter()` method just after creating the node, and before spinning:

```
auto node = std::make_shared<ResetCounterClientNode>();
node->callResetCounter(20);
rclcpp::spin(node);
```

Now, create a new executable in `CMakeLists.txt`. Build the package, open a few Terminals, and start the `number_publisher` and `number_counter` nodes. Then, start the `reset_counter_client` node to try the service communication.

Now that you've written the code for both the service server and client, let's explore what you can do with the ROS 2 tools. For services with a simple interface, you will be able to test them directly from the Terminal, even before writing the code for a client.

Additional tools to handle services

We've already used the `ros2` command-line tool a lot in this book. With this tool, each core ROS 2 concept gets additional functionalities in the Terminal. This is no exception for services.

We're now going to explore `ros2 service` a bit more so that we can introspect services and send a request from the Terminal. We will also learn how to change a service name at runtime (`ros2 run`).

To see all commands for ROS 2 services, type `ros2 service -h`.

Listing and introspecting services

First, `rqt_graph` does not support services (yet—there are plans to maybe add this in a future ROS 2 distribution), so we won't use it here. We will only use the `ros2` command-line tool.

Stop all nodes and start the `number_counter` node. Then, to list all services, run the following command:

```
$ ros2 service list
/number_counter/describe_parameters
/number_counter/get_parameter_types
/number_counter/get_parameters
/number_publisher/get_type_description
/number_counter/list_parameters
/number_counter/set_parameters
/number_counter/set_parameters_atomically
/reset_counter
```

For each node you start, you will get seven additional services, mostly related to parameters. You can ignore those. If you look at the list, apart from those seven services, we can retrieve our `/reset_counter` service.

> **Note**
> Note that there is an additional leading slash in front of the service name. Service names follow the same rules as nodes and topics. If you don't provide any namespace (for example, `/abc/reset_counter`), you're in the "global" namespace, and a slash is added at the beginning.

Once you have the service name you want, you can get the service interface with `ros2 service type <service_name>`:

```
$ ros2 service type /reset_counter
my_robot_interfaces/srv/ResetCounter
```

From this, you can see the details inside the interface:

```
$ ros2 interface show my_robot_interfaces/srv/ResetCounter
int64 reset_value
---
bool success
string message
```

This process is extremely useful when you need to create a service client for an existing server. There's no need to even read the server code—you can get all the information you need from the Terminal.

Sending a service request

To test a service server, you usually have to write a service client.

Good news: instead of writing a client, you can call the service from the Terminal directly. This can save you some development time.

First, you must know the service name and interface. Then, use the `ros2 call <service_name> <interface_name> "<request_in_json>"` command. Let's try this with our `reset_counter` service:

```
$ ros2 service call /reset_counter \
my_robot_interfaces/srv/ResetCounter {reset_value: 7}"
waiting for service to become available...
requester: making request: my_robot_interfaces.srv.ResetCounter_Request(reset_value=7)

response:
my_robot_interfaces.srv.ResetCounter_Response(success=True, message='Success')
```

You can see the request being sent, followed by the response. Then, the command exits. This is practical and in this case, we save a lot of time.

We can also easily test the different cases. For example, let's send a negative value for the reset number:

```
$ ros2 service call /reset_counter \
my_robot_interfaces/srv/ResetCounter "{reset_value: -7}"
waiting for service to become available...
requester: making request: my_robot_interfaces.srv.ResetCounter_Request(reset_value=-7)

response:
my_robot_interfaces.srv.ResetCounter_Response(success=False, message='Cannot reset counter to a negative value')
```

With this example, it's quite easy as the request is very simple (only one integer number). For more complex service requests that contain lots of nested fields and arrays, writing the full request in the Terminal can become quite cumbersome, and you will spend a lot of time trying to get it right.

So, for simple interfaces, use `ros2 service call` to try the service first. For more complex interfaces, you'll have to write a client code first. This isn't really a problem: you can use the code we used for `ResetCounterClientNode` as a template for any other client. In the end, both methods allow you to test a service server quite quickly.

Changing a service name at runtime

When you start a node with `ros2 run`, you can change the node name and any topic name inside the node. You can do the same for services.

As a reminder, for any additional argument you pass after `ros2 run`, add `--ros-args`, but only once.

Then, to rename a service, add `-r` followed by `<service_name>:=<new_service_name>`.

For example, let's rename the `reset_counter` service to `reset_counter1` when we start the `number_counter` node:

```
$ ros2 run my_py_pkg number_counter --ros-args -r \
reset_counter:=reset_counter1
```

Now, let's verify this with `ros2 service list`:

```
$ ros2 service list
# Some other services
/reset_counter1
```

The service name is now `/reset_counter1`. If we start a node with a service client, we need to modify the name as well; otherwise, the nodes won't be able to communicate with each other:

```
$ ros2 run my_py_pkg reset_counter_client --ros-args -r \
reset_counter:=reset_counter1
```

Doing this is quite useful, especially when you want to run several nodes (written by yourself or others) that use a slightly different service name, or that are in different namespaces.

You are now able to write a service server/client and introspect/test them from the Terminal. Before moving on to the next chapter, let's practice more with an additional challenge.

Service challenge – client and server

With this new challenge, you will practice everything that was covered in this chapter: custom service interfaces, service servers, and service clients.

We will use the `turtle_controller` node we wrote in the previous chapter's challenge as a starting point. We won't create a new node here; instead, we will improve the existing code. You can either start from the code you wrote or from the code I provided in the `ch5` folder of this book's GitHub repository.

As always, I will explain what you need to do to complete the challenge, and then detail the most important points for the Python solution. You can find the complete solution code on GitHub for both Python and C++.

Challenge

This challenge is divided into two parts. I suggest following them in order.

Challenge 1 – service client

So far, our `turtle_controller` node is subscribing to the `/turtle1/pose` topic. In the subscriber callback, we send a velocity command to the `/turtle1/cmd_vel` topic.

The result of this is the turtle drawing a circle on the screen, with a different velocity depending on if it is on the right or left of the screen.

What we want to do now is change the color of the pen, depending on where the turtle is. If the turtle is on the right of the screen, we want the pen color to be red. On the left, the color should be green.

To do that, we will need to add a service client in the node so that we can call the service to change the pen's color in the `turtlesim` node (I won't give you the service name—that's part of the challenge).

Here are the steps you can take to get started:

1. Start the `turtlesim` node and use the `ros2 service` command line to find which service to call, as well as what interface to import (optional: at that point, you can also test the service with `ros2 service call`, directly from the Terminal).
2. In the `turtle_controller` node, add a service client for that service.
3. Create a method that will call the service.
4. Call this method from the existing subscriber callback. After you publish the new command velocity, check whether the turtle is on the right or left of the screen. When the turtle switches to a different side, call the service with the updated color.

Challenge 2 – custom interface and service server

Once you're done with the first challenge, try this one. This time, you'll practice on the server side of services.

Here, we want to allow the `turtle_controller` node to activate or deactivate the turtle (meaning to start or stop the turtle), depending on an external request. For that, we will create a service server.

Here are the steps you can take to get started:

1. Define a service name and interface for that service.
2. If no existing interface matches your needs, you will need to create and build a new one (hint: that's what we will do here).
3. In the `turtle_controller` node, add a service server and a callback, in which you activate or deactivate the turtle. Tip: you can use a simple boolean attribute in the class to store the activated state for the turtle.
4. If the turtle is activated, then in the subscriber callback, you can keep sending additional velocity commands. If it is not activated, you don't send any commands.

With those instructions, you should be able to get started. Taking the time to do this exercise is probably the best investment you can make to learn ROS faster.

Solution

Let's start with the first challenge.

Challenge 1

For this challenge, we are on the client side, which means that we need to find out which service we need to call. I will do a quick recap of the steps for finding the service name and interface.

Start the `turtlesim` node and list all services. You should see a `/turtle1/set_pen` service with `ros2 service list`.

Now, get the type for this service and see what's inside the interface:

```
$ ros2 service type /turtle1/set_pen
turtlesim/srv/SetPen
$ ros2 interface show turtlesim/srv/SetPen
uint8 r
uint8 g
uint8 b
uint8 width
uint8 off
---
```

In the service request, we can send an (r,g,b) value (red, green, blue). There are also `width` and `off` attributes, but we won't use them.

At this point, before you even write the code for the client, you can try the service from the Terminal:

```
$ ros2 service call /turtle1/set_pen turtlesim/srv/SetPen \
"{r: 255, g: 0, b: 0}"
```

Then, execute `ros2 run turtlesim turtle_teleop_key` and move the turtle around. You'll see that the pen is now using a red color.

Back to the code, inside the `turtle_controller.py` file, import the interface:

```
from turtlesim.srv import SetPen
```

Since we've already added the dependency for `turtlesim` in the `package.xml` file of the `turtle_controller` package (in the previous chapter), there's no need to do it again.

Then, create the service client in the constructor:

```
self.set_pen_client_ = self.create_client(SetPen, "/turtle1/set_pen")
```

Write the method that will call the service, as well as the callback for the response:

```
def call_set_pen(self, r, g, b):
    while not self.set_pen_client_.wait_for_service(1.0):
        self.get_logger().warn("Waiting for service...")
    request = SetPen.Request()
    request.r = r
    request.g = g
    request.b = b
    future = self.set_pen_client_.call_async(request)
    future.add_done_callback(
self.callback_set_pen_response)

def callback_set_pen_response(self, future):
    self.get_logger().info("Successfully changed pen color")
```

We only send the r, g, and b parts of the request. The other values (width and off) will be kept as-is.

As you can see, in the callback for the response, we don't check what's inside the response since the response is empty (it exists but it doesn't contain a field).

The only thing we need to do now is call this new call_set_pen() method. We will do that from within the subscriber callback since this is where we have access to the X position of the turtle.

Inside the callback_pose() method, and after the code to publish on the topic, add the code to handle the pen color:

```
if pose.x > 5.5 and self.previous_x_ <= 5.5:
    self.previous_x_ = pose.x
    self.get_logger().info("Set color to red.")
    self.call_set_pen(255, 0, 0)
elif pose.x <= 5.5 and self.previous_x_ > 5.5:
    self.previous_x_ = pose.x
    self.get_logger().info("Set color to green.")
    self.call_set_pen(0, 255, 0)
```

If the turtle is on the right, we set the color to red (255, 0, 0), and if it's on the left, we set the color to green (0, 255, 0).

On top of that, we also define a new attribute in the constructor so that we can keep track of the previous X coordinate:

```
self.previous_x_ = 0.0
```

We use this to only call the service when the turtle switches from one side to the other. Why do we do that? We could send a service request every time, even if the color would be the same as the previous one. Why "optimize" the code?

The reason is that the `callback_pose()` method will be called a lot. Check the frequency for the `/turtle1/pose` topic in the Terminal:

```
$ ros2 topic hz /turtle1/pose
average rate: 62.515
```

This means that we execute `callback_pose()` about 62 times per second. This is not really a problem. We also publish on the `/turtle1/cmd_vel` topic at 62 Hz. Again, that's not a problem. Publishers and subscribers can sustain a high frequency (with a bigger message size, this could become complicated, but here, the messages are really small).

Now, what if we send a request to a service 62 times per second? This is where the problem is. Services are not made for high-frequency requests, and this could seriously affect the performance of the application. Also, if you find yourself having to call a service at 62 Hz, then you probably have a design problem, and you either need to modify your code to reduce the frequency or use a publish/subscribe mechanism instead.

So, what we do in the code is make sure we only call the service when it's needed—that is, when the turtle switches from one side to the other.

The code is now complete! At this point, you can build your `turtle_controller` package again (unless you have already built it with `--symlink-install`), source the environment, and then start both the `turtlesim` and `turtle_controller` nodes to see the result.

Challenge 2

Now, we want to add a service server inside our node so that we can activate or deactivate the turtle. Since we're defining the server, we need to come up with a name and an interface:

- **Name**: Let's use `activate_turtle`. We'll start with a verb and try to make the name as explicit as possible.
- **Interface**: If you look at existing interfaces, we could use the `SetBool` service from `example_interfaces`. It contains a boolean in the request and a string in the response. However, as stated previously, the best practice is to avoid using the `std_srvs` and `example_interfaces` packages if your application is any serious. So, in this case, we'll create our own interface.

Let's create a new interface for our service. This will be quite quick and easy as we already have the `my_robot_interfaces` package fully configured.

Inside the `srv` folder of the `my_robot_interfaces` package, create a new service file named `ActivateTurtle.srv`. In this file, write the service definition:

```
bool activate
---
string message
```

This is all we need in the request: a boolean to activate or deactivate the turtle. We also added a string in the response so that we get to know what happened, but you could also decide to have an empty response.

After this, add the interface to the `CMakeLists.txt` file of the `my_robot_interfaces` package, and build the package. Source the environment, and make sure you can see the interface with:

```
ros2 interface show my_robot_interfaces/srv/ActivateTurtle
```

Now, let's go back to the `turtle_controller` package.

Since we will have a dependency on `my_robot_interfaces`, add a new line to the `package.xml` file of the `turtle_controller` package:

```
<depend>my_robot_interfaces</depend>
```

Now, it's time to write the code inside `turtle_controller.py`. Import the interface:

```
from my_robot_interfaces.srv import ActivateTurtle
```

In the constructor, add a boolean flag to keep track of the activated status for the turtle, and create a new service server:

```
self.is_active_ = True
self.activate_turtle_service_ = self.create_service(ActivateTurtle,
"activate_turtle", self.callback_activate_turtle)
```

Implement the callback method for that service:

```
def callback_activate_turtle(self, request: ActivateTurtle.Request,
response: ActivateTurtle.Response):
    self.is_active_ = request.activate
    if request.activate:
        response.message = "Starting the turtle"
    else:
        response.message = "Stopping the turtle"
    return response
```

What we do is simple—we just set the `is_active_` boolean with the value we get from the boolean in the request. Now, whenever you call this service, the `is_active_` boolean will be updated with the value you send.

The last step, to make the turtle start or stop when activated or deactivated, is to modify the code inside the `callback_pose()` method:

```
def callback_pose(self, pose: Pose):
    if self.is_active_:
        # Entire code for the callback, inside the "if"
```

This way, we only publish a new command velocity if the turtle is activated. If not, we publish nothing. Also, the service request will only work when the turtle is activated.

To try this new service, start the `turtlesim` node and `turtle_controller` nodes. In a third Terminal, send a service request with the `ros2` command-line tool. Here's an example:

```
$ ros2 service call /activate_turtle \
  my_robot_interfaces/srv/ActivateTurtle "{activate: false}"
```

This should make the turtle stop. You can send a request again, this time with "`{activate: true}`", which should make the turtle move again.

That's the end of this challenge on services. If you managed to finish this challenge by yourself, you have a good understanding of services. No worries if you couldn't do it without having to look at the solution. Come back to it in a few days and see if you can solve the challenge again.

Summary

In this chapter, you worked on ROS 2 services, which is another ROS 2 communication you can use alongside topics.

With services, nodes can talk to each other using a client/server type of communication. Only one server can exist for a service, but you can send multiple requests from several clients.

You can implement service servers and clients directly in your nodes using `rclpy` for Python and `rclcpp` for C++.

To write a service server, you must do the following:

1. As the name and interface are defined by the server, you have to choose them here. As a best practice, use a verb as the first word in the name.
2. Import the interface in your code and create the service server in the constructor.
3. Add a callback method to process any received request.

When choosing a service interface, if you can't find an existing one that perfectly matches what you need, then you have to create and build your own. To do that, you must do the following:

1. Create and set up a package dedicated to interfaces. If you already have one for your application, use it.
2. Add the new service interface to the package and build it.
3. Now, you can use this interface in your service server.

To write a service client, do the following:

1. If you're writing a client, it means that there is an existing server on the other side. Find which name and interface you need to use.
2. Import the interface into your code and create the service client in the constructor.
3. Create a method to call the service. In this method, you send the request asynchronously, and then register a callback to process the response.
4. You can call the service from anywhere in your code.

With the `ros2 service` command line, you can introspect the services in your nodes and see what interface they're using.

To try a service server, you can either write a corresponding service client inside another node or, if the request is simple, call the service directly from the Terminal with `ros2 service call`.

You have now seen the two most common ROS 2 communication types: topics and services. In the next chapter, we will work with the third and last one: actions.

7
Actions – When Services Are Not Enough

In this chapter, we will explore the third communication type in ROS 2: actions. To understand actions, you need to have read the previous chapters on nodes, topics, and services.

Before we begin, I want to alert you that this chapter covers more advanced material compared to what we encountered previously and what's to come.

If you already have some level of expertise, this chapter will satisfy you as it will give you a full overview of all three ROS 2 communication types. However, if you're just getting started with ROS with zero experience, it might be a bit too much for you right now. This is OK, and topics/services are more than enough to get started with ROS 2. You can skip this chapter (which is independent of future chapters) for now and continue with parameters and launch files. It might be a good idea to come back to it at a later stage after you've built more confidence by working on ROS 2 projects.

Throughout this chapter, you will understand why you need actions and how they work by going through an example that we will build step by step. Then, you will write the code to make two nodes communicate with each other. We will use the code in the ch6 folder (in this book's GitHub repository: https://github.com/PacktPublishing/ROS-2-from-Scratch) as a starting point. You can find the final code in the ch7 folder.

By the end of this chapter, you will be able to write an action server and client and take advantage of all action features, such as feedback and cancel mechanisms.

Even though topics and services are more than enough to get started, ROS 2 actions are important as they help you take your code to the next level and implement more complex behaviors in your robotics applications.

In this chapter, we will cover the following topics:

- What is a ROS 2 action?
- Creating a custom action interface

- Writing an action server
- Writing an action client
- Taking advantage of all the action mechanisms
- Additional tools to handle actions

What is a ROS 2 action?

To understand ROS 2 actions, we need to understand why we need them. That's what we will focus on first. After that, I will explain how actions work through a real-life example.

You quickly discovered actions in *Chapter 3* by running some existing nodes and command-line tools. The intuition you built there will help you better understand the concepts in this chapter.

Let's dive in and see why and when actions could be needed in a ROS 2 application.

Why actions?

So far, we've looked at two forms of communication in ROS 2: topics and services.

Topics are used by nodes to send and receive messages. Publishers will publish data on a topic, and subscribers will subscribe to the topic to receive the data. Thus, topics are perfect for sending data streams in your application.

Services are used for client/server interactions between nodes. The client sends a request to the server, after which the server executes or computes something and returns a response to the client.

So, that should be all, right? What else could we need?

In its early days, ROS started with only topics and services. However, ROS developers quickly realized that something was missing for some robotics applications. Let's see that with an example.

Imagine that you have a mobile robot with two wheels. First, you would create a node that's responsible for controlling the wheels. This node would also be able to receive commands, such as `Move to (x, y) coordinates`. Those commands would be transformed into a velocity to apply to the wheels. However, you would also like to be able to get notified when the robot has finished moving.

With what we know so far, ROS 2 services seem to be a good option. In this server node, you could implement a `/move_robot` service that will receive coordinates from a client. Once the command is received, the controller starts to move the wheels. Then, when the robot has reached its destination, the server returns a response to the client.

To complete the communication, we must add a service client to another node. The client will send a request to the server with the (x,y) coordinates to reach. When the server returns the response, we know that the robot has finished moving—either successfully by reaching the destination or something prevented it and we get an error:

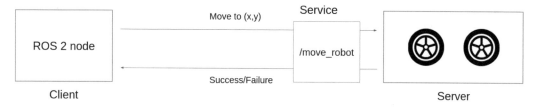

Figure 7.1 – Using a service to control a two-wheeled robot

What's wrong with that?

Well, moving a physical part of a robot in space can take some time. It could be a fraction of a second in some cases, but also maybe a few seconds, or even a few minutes. The point is that the service execution could take a significant amount of time.

With that said, while the robot is moving, there are a few things you may want to do, and those things are missing when using services:

- Since the execution is taking some time, it would be nice to get some feedback from the server. With a service, the client has no idea of what's happening on the server side. So, the client is completely blind and needs to wait for the response to get some information.

- How can you cancel the current execution? That would seem a reasonable feature to have. After you start the execution on the server side, the client may want to cancel it. For example, let's say the client node is also monitoring the environment with a camera. If an obstacle is detected, the client could ask the server to stop the execution. With what we have for now, the client can't do anything but wait for the server to finish the execution.

- Here's the last point for now, although we could find more: how could the server correctly handle multiple requests? Let's say you have two or more clients, each one sending a different request. How can you possibly choose between those requests on the server? How can the server refuse to execute a request, or choose to replace a request with a new one, without finishing the first request? Or, in another scenario, how can the server handle multiple requests at the same time? As an analogy, when you download files on your computer, the computer isn't stuck with just one file. It can download multiple files at the same time. You can even decide to cancel one download while the others are still running.

Coming back to our example, you can see that a simple service is not enough. For this use case, we need more functionalities. What we could do is implement additional services, such as one to cancel a request. We could also add a new topic to publish some feedback about where the robot is during the execution.

There's good news—you don't have to do this. All these problems are solved by ROS 2 actions. The feedback mechanism, cancel mechanism, and other functionalities are also implemented directly in actions.

To conclude, services are perfect for client/server communication, but only if the action/computation is quick to execute. If the execution could take some time, and you want additional features such as feedback or cancellation, then actions are what you need.

Now that you know why we need actions, it's time to understand how they work.

How do actions work?

Let's use the previous example, this time using a ROS 2 action instead of a service. I will show you how actions work at a high level, with the different interactions between the client and the server. Later in this chapter, we will dive into the code and see the implementation details.

We will use two nodes: one containing an **Action client**, and the other containing an **Action server** (this is the one responsible for controlling the wheels).

To understand how actions work, let's follow the execution flow for one action:

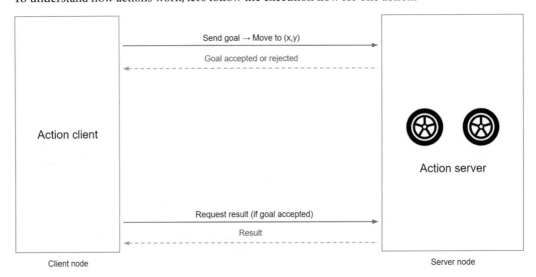

Figure 7.2 – Execution flow for a ROS 2 action

Here are the steps for this flow:

1. The **Action client** will start the communication by sending a request to the **Action server**. For actions, the *request* is named *goal*. Hence, we won't talk about requests here, but about **goals**. Here, the goal can be a (x, y) coordinate to reach.

2. The **Action server** receives the goal and decides to accept or reject it. The client immediately receives this response from the server. If the goal has been rejected, then the communication ends.
3. If the goal is accepted, the server can start to process it and execute the corresponding action. With this example, the **Server node** will make the robot move.
4. As soon as the client knows that the goal has been accepted, it will send a request to get the **Result** and wait for it (asynchronously, by registering a callback). For services, we talk about a *response*. For actions, this will be a *result*.
5. When the server is done executing the goal (either successfully or not), it will send the **Result** to the client. With this example, the result could be the final reached (x, y) coordinates.
6. The client receives the **Result**, after which communication ends.

This is how an action works, with a minimal set of functionalities. From the server side, a goal is received, accepted or rejected, then executed, and the result is returned. From the client side, a goal is sent, and if accepted, a request is sent and the result is received from the server.

On top of that, you can add extra functionalities, all of which are optional. Here are the additional mechanisms for actions:

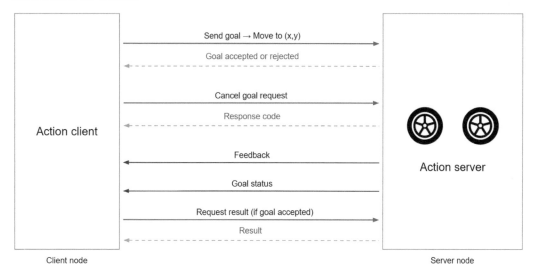

Figure 7.3 – Action with all communication mechanisms

Let's take a closer look:

- **Feedback**: The server, while executing the goal, can send some feedback to the client. With this example, the feedback could be the current coordinates for the robot or even a completion rate. Thus, the client can know what's happening during the goal's execution.

- **Cancel**: After the goal has been accepted by the server and the goal is being executed on the server side, the client can decide to cancel that goal. To do so, it will send a cancel request that must be approved by the server. If the cancel request is accepted, then the server will try to finish the execution. So, in this example, it could make the robot stop. In the end, the server will still return a result to the client, whether the goal was successful, failed, or canceled.

- **Goal status**: This is not so important for you as it's an internal mechanism for actions that you will not use directly in your code (I just added it here for completeness). Each goal will get a state machine, with states such as *accepted*, *executing*, and others. With each change of state for a goal, the server will notify the client.

With this, you have seen all possible communication mechanisms that can be implemented within actions.

Note that in the preceding figure, some communications are represented with a red line, while others are presented with a green line. Behind the scenes, actions just use topics and services. Even if an action is a ROS 2 communication on its own, the underlying code is using the two other communication types. Here, red lines represent services, and green lines represent topics.

Thus, within an action, you have three services (send goal, cancel goal, and receive result) and two topics (feedback and goal status). The good news is that you don't have to create those topics and services yourself—they are already implemented in the action mechanism. All you have to do is use the action client and server functionalities from the ROS 2 libraries.

To create an action, you will need to give it a **name** (for example, `move_robot`) so that the client knows where to send the goal. You will also need to use an **interface (goal, result, feedback)**.

One additional thing to note is that there can be only one action server. Just as for services, you can't have two servers using the same name. On the other hand, you can have multiple action clients. Each client can also send multiple goals; that's not a problem.

Wrapping things up

On top of topics and services, you can use actions to make your nodes communicate with each other. Now, when should you use topics, services, or actions?

You should use *topics* when you want to send data streams between nodes. With topics, there's no response. For example, this can work for publishing sensor data or sending a stream of commands to another node if you don't need any confirmation.

Services are perfect when you want client/server communication, but also if the action to execute is very quick, such as a computation or a simple action, such as switching on an LED.

Finally, you will use *actions* for anything that needs client/server communication and may take some time to execute, as well as when you also want to have mechanisms such as feedback and cancel.

Here are some important points about how actions work:

- An action is defined by a name and an interface.
- The name of an action follows the same rules as for topics and services. It must start with a letter and can be followed by other letters, numbers, underscores, tildes, and slashes. Also, as the action is *doing* something, the best practice is to start the name with a verb.
- The interface contains three things: a goal, a result, and feedback. Both the client and server must use the same interface.
- An action server can only exist once, but you can send multiple goals from one or multiple action clients.
- Action clients aren't aware of the node containing the server. They just know they have to use the action name and interface to reach the server.

To implement an action communication, you will need to do the following at the very least:

- Send a goal from the client to the server.
- Accept (or not) the goal and execute it on the server.
- Once the goal is finished, return a result from the server to the client.

The following are some optional features you can add:

- Send some execution feedback from the server to the client while the goal is being executed.
- Allow the client to send a cancel request to the server. If accepted, finish the goal execution on the server side.

To write action servers and clients in your code, you must use the action functionality from the `rclpy.action` and `rclcpp_action` libraries.

At this point, we can start writing some code. If you're still a bit confused, don't worry—actions are quite complex to grasp initially. They contain lots of different mechanisms. Everything will make more sense as we create an action and write the client and server code.

Since we can't test a client without a server, we will, as we did for services, start with the server side. To create a server, we need an action interface, so that will be our starting point.

Creating a custom action interface

To create an action interface, we first need to clearly define what we need to achieve with the action. Then, we can add the interface to the `my_robot_interfaces` package (in this section, we will continue using the packages we created in the previous chapters).

Defining the application and the interface we need

In the application that we will write in this chapter, the action server will be responsible for counting until a given number, with a delay between each count, so that we can simulate that the action takes some time and doesn't return immediately. The client will have to send a number to the server so that the server can start to count. When the server finishes, it will send the result (last reached number) back to the client.

For example, let's say the client sends the number 5 to the server, and there's a delay of 0.5 seconds. The server will start to count from 0, up to 5, and wait 0.5 seconds between each iteration. When finishing, the server will return 5 if it could count until the end, or the last reached number if the execution finished sooner (the goal was canceled, or any other reason that could make the server stop the goal). In addition to that, we will add some feedback about the current count while the server is executing the goal.

Before we write any code, we need to know what interface to use for the action. From the previous paragraph, we can see that we need the following:

- **Goal**: An integer for the target number and a float number for the delay
- **Result**: An integer for the last reached number
- **Feedback**: An integer for the current count

For topics and services, you must first check whether you can find an existing interface that matches your needs as there are already a lot of them you can use without having to create a new one.

For actions, you could try to do the same, but there aren't as many existing action interfaces. Actions are a bit more complex than the other communication types, so you would need to find an interface that matches the goal, result, and feedback for your application exactly. The probability of that is very low as each action will be quite different. Thus, for actions, we won't try to find existing interfaces and create a custom one directly.

Creating a new action interface

The process of creating an action interface will be the same as for topic and service interfaces. We will follow a similar approach.

First, you need to create and configure a package dedicated to interfaces. We did that in *Chapter 5*, in the *Creating a custom interface for a topic* section, with the `my_robot_interfaces` package. You can reuse this package to add your action interfaces. If you don't have it, go back and configure it first, then continue with the following steps.

In this package, we already have `msg` and `srv` folders for topic and service interfaces, respectively. We will add a third folder, named `action`, for—as you may have guessed—action interfaces:

```
$ cd ros2_ws/src/my_robot_interfaces/
$ mkdir action
```

In this new folder, you will place all the action interfaces specific to your robot or application.

Now, create a new file for your action. Here are the rules you must follow regarding the filename:

- Use UpperCamelCase—for example, `CountUntil`.
- Don't write `Action` or `Interface` in the name as this will add unnecessary redundancy.
- Use `.action` for the file extension.
- As a best practice, use a verb in the interface's name—for example, `NavigateToPosition`, `OpenDoor`, `PickObjectFromTable`, or `FetchDrinkFromFridge`. Actions, just like services, are about performing an action or computation (which can take some time), so by using a verb, you make it very clear what the action is doing.

Since we want to count until a given number, let's call the interface `CountUntil`:

```
$ cd ~/ros2_ws/src/my_robot_interfaces/action/
$ touch CountUntil.action
```

You can write the definition for the action in this file. Since we have three different parts (goal, result, and feedback), we need to separate them. You must add three dashes (---) between the goal and the result, and another three dashes between the result and the feedback.

Even if you don't want to send any feedback, or if the result is empty, you still have to add the two separations with three dashes (---). A very simple action definition with nothing in the result and feedback would look like this:

```
int64 goal_number
---
---
```

For the goal, result, and feedback, you can use the following:

- Built-in types (`bool`, `byte`, `int64`, and so on).
- Existing message interfaces. For example, the goal of the action could contain `geometry_msgs/Twist`.

> **Note**
>
> You can't include an action or service definition inside an action definition. You can only include messages (topic definition) inside the goal, result, or feedback. Those three parts can be seen as three independent messages.

Since we are creating a rather simple application, we will only use built-in types here:

```
# Goal
int64 target_number
float64 delay
---
# Result
int64 reached_number
---
# Feedback
int64 current_number
```

As for topic and service interfaces, all fields inside the definition must follow the `snake_case` convention (use underscores between words, all letters must be lowercase, and no spaces).

I've also added comments to specify which part is the goal, result, and feedback. You don't need to do this—I only did it for your first action definition so that you don't get confused. Often, people make mistakes regarding the order and put the feedback before the result, which can lead to hard-to-debug errors later. The order is goal, result, and then feedback.

Now that we've written our interface, we need to build it so that we can use it in our code. Go back to the `CMakeLists.txt` file of the `my_robot_interfaces` package. Since the package has already been configured, we just need to do one thing: add the relative path to the interface on a new line inside the `rosidl_generate_interfaces()` function. Don't use any commas between the lines:

```
rosidl_generate_interfaces(${PROJECT_NAME}
  "msg/HardwareStatus.msg"
  "srv/ResetCounter.srv"
  "srv/ActivateTurtle.srv"
  "action/CountUntil.action"
)
```

After this, save all files and build the `my_robot_interfaces` package:

```
$ colcon build --packages-select my_robot_interfaces
```

Once built, source the environment. You should be able to find your new interface:

```
$ ros2 interface show my_robot_interfaces/action/CountUntil
# Action interface definition here
```

If you see the action definition, you know that your action interface has been successfully built, and you can now use it in your code. That's what we will do, starting with the action server for our application.

Writing an action server

In this section, you'll write your first action server. In this server, we will be able to receive goals. When a goal is received, we will decide whether to accept or reject it. If it's accepted, we will execute the goal. For this application, executing the goal means we will start to count from zero to the target number and wait for the provided delay between each iteration. Once the goal has been executed, we will return a result to the client.

That's what we will implement in the code, starting with Python and then C++. In this section, we start only with the minimum functionalities for the action communication to work correctly. We will add the feedback and cancel mechanisms later. Since actions are a bit more complex than topics and services, let's start simple and go step by step.

For a better learning experience, make sure you use the GitHub code while following along as I will not necessarily display all lines in this chapter, only the important ones. The code for this section is located in the count_until_server_minimal file (with .py or .cpp appended at the end). We won't use the number_publisher and number_counter nodes here.

Before we write any code for the server, we need to choose a name and interface for our action. Since we want to count until a given number, we will name the action count_until, and we will use the CountUntil interface we've just created.

We now have everything we need to start writing the Python code.

Writing a Python action server

You will need to write your action server inside a node. Create a new file named count_until_server_minimal.py inside the my_py_pkg package (along with the other Python files). Make this file executable.

Importing the interface and creating the server

Let's start by setting up the action server.

First, we must import a bunch of libraries and classes that we will need in the code:

```
import rclpy
import time
from rclpy.node import Node
from rclpy.action import ActionServer, GoalResponse
from rclpy.action.server import ServerGoalHandle
```

Unlike topics and services, the action server is not directly included in the `Node` class. So, we need to import the `ActionServer` class from `rclpy.action`.

After this, you must also import the interface for the action:

```
from my_robot_interfaces.action import CountUntil
```

When you import an interface from another package, make sure to add the dependency to `my_robot_interfaces` in the `package.xml` file of `my_py_pkg` (you should have already done this if you've been following along):

```
<depend>my_robot_interfaces</depend>
```

Going back to the `count_until_server_minimal.py` file, let's create the action server in the node's constructor (as stated in the introduction to this section, I'll only display the important and relevant snippets; the full constructor code is available on GitHub):

```
self.count_until_server_ = ActionServer(
        self,
        CountUntil,
        "count_until",
        goal_callback=self.goal_callback,
        execute_callback=self.execute_callback)
```

To create an action server with Python, you must use the `ActionServer` class we imported previously. Provide the following arguments:

- **Action node**: The node to link the action server to. For topics and services, we started with `self.create...()`. Here, it's a bit different: the object (`self`) is provided as the first argument.
- **Action interface**: We use the `CountUntil` interface we've imported.
- **Action name**: Since we're writing the code for the server, we're creating the action here. This is where you will choose the action name that all clients will have to use to send goals. As seen previously, we will use `count_until`.
- **Goal callback**: When a goal is received, it will be processed inside this callback.
- **Execute callback**: If the goal has been accepted in the goal callback, then it will be processed in the execute callback. This is where you will execute the action.

We specified two callback methods when creating the action server. When the node spins, the action server will be in *waiting mode*. As soon as a goal is received, the node will trigger the goal callback, and then the execute callback if needed. Let's implement those callbacks.

Accepting or rejecting a goal

The action server can now receive goals. We need to decide whether to accept or reject them.

Let's start writing the goal callback, which is the first method to be called whenever a goal is received by the server:

```
def goal_callback(self, goal_request: CountUntil.Goal):
    self.get_logger().info("Received a goal")
    if goal_request.target_number <= 0:
        self.get_logger().warn("Rejecting the goal, target number must be positive")
        return GoalResponse.REJECT
    self.get_logger().info("Accepting the goal")
    return GoalResponse.ACCEPT
```

In this callback, we receive the goal that was sent by the client (it's of the CountUntil.Goal type).

> **Note**
> An action interface contains a goal, a result, and feedback. You get one class for each message: CountUntil.Goal, CountUntil.Result, and CountUntil.Feedback. We will use all three in this chapter.

The best practice is to validate the data you receive whenever you write the code for a server. For this application, let's say we want to only accept positive target numbers. If the number is negative, we reject the goal.

After validating the data, you need to return either GoalResponse.ACCEPT or GoalResponse.REJECT to accept or reject the goal, respectively. The client will be notified immediately of that decision. Then, if the goal is rejected, nothing more happens on the server side. If the goal is accepted, the execute callback will be triggered.

Executing the goal

Let's implement the execute callback. Here's the beginning of the code:

```
def execute_callback(self, goal_handle: ServerGoalHandle):
    target_number = goal_handle.request.target_number
    delay = goal_handle.request.delay
    result = CountUntil.Result()
    counter = 0
```

In this callback, you get what's called a goal handle, which is of the `ServerGoalHandle` type. I've made the argument type explicit so that we can get auto-completion with VS Code. This goal handle contains the goal information, but you can also use it to set the goal's final state, which we will see in a minute.

The first thing you must typically do is extract the data from the goal. Here, we get the target number and delay that we will use when executing the action. Then, we initialize a few things: the result from the `CountUntil.Result` class, and a counter starting at 0.

With this, we can start to execute the goal:

```
self.get_logger().info("Executing the goal")
for i in range (target_number):
    counter += 1
    self.get_logger().info(str(counter))
    time.sleep(delay)
```

This part of the code will be different every time as it depends entirely on your application. Here, we're incrementing the counter until the target number is reached, with a delay between each iteration.

The point of using a delay here is just to make this method take more time so that we can simulate the behavior of an action. If we wanted to count as fast as possible, without any delay, we could have used a service since the action would finish almost immediately.

Once the execution is finished, we need to do two things—set a final state for the goal and return a result to the client:

```
goal_handle.succeed()
result.reached_number = counter
return result
```

During the execution of the action, the goal is in the *executing* state. When finishing the execution, you need to make it transition into a final state.

In this case, since everything went smoothly and we didn't expect any problems during the execution, we set the goal to *succeeded* by using the `succeed()` method on the goal handle. If, for example, your action was responsible for moving the wheels of a robot, and if the communication with the wheels is lost during the execution, you would stop the action and set the goal to *aborted* with the `abort()` method. The last possible state is *canceled*, which we will see a bit later in this chapter.

We've now written the minimal code for the action server to work properly. Before we write an action client, let's switch to C++. If you only want to follow the Python explanations, then go ahead and skip the next section.

Writing a C++ action server

The code logic for C++ actions is very similar to Python, but there are quite a few specificities about the syntax. We will focus mostly on those differences. Also, as the code starts to become quite large, I will not necessarily display the full code, only the important parts for comprehension. Make sure you take a look at this book's GitHub repository to see the full code.

Importing the interface and creating the server

Let's start by setting up the action server. First, create a new file named count_until_server_minimal.cpp in the src directory of your my_cpp_pkg package.

Open the file and start by adding the necessary includes:

```
#include "rclcpp/rclcpp.hpp"
#include "rclcpp_action/rclcpp_action.hpp"
#include "my_robot_interfaces/action/count_until.hpp"
```

As you can see, the action library is not a sub-library of rclcpp—it's a completely independent one from a different package: rclcpp_action.

For each new package we use, we need to add the dependency to the package.xml file of the my_cpp_pkg package:

```
<depend>my_robot_interfaces</depend>
<depend>rclcpp_action</depend>
```

You will also need to specify those dependencies in the CMakeLists.txt file:

```
find_package(my_robot_interfaces REQUIRED)
find_package(rclcpp_action REQUIRED)
```

Finally, when you create your executable, don't forget to add both dependencies to the ament_target_dependencies() function:

```
add_executable(count_until_server src/count_until_server_minimal.cpp)
ament_target_dependencies(count_until_server rclcpp rclcpp_action my_robot_interfaces)
```

Back to the count_until_server_minimal.cpp file, we add a few using lines to simplify the code (you can find those lines at the top of the file, under the #include lines). After that, you can add an action server to your class as a private attribute:

```
rclcpp_action::Server<CountUntil>::SharedPtr count_until_server_;
```

Once again, we're going to use a shared pointer to keep the object.

Then, in the constructor, you can create the action server:

```cpp
count_until_server_ = rclcpp_action::create_server<CountUntil>(
    this,
    "count_until",
    std::bind(&CountUntilServerNode::goalCallback, this, _1, _2),
    std::bind(&CountUntilServerNode::cancelCallback, this, _1),
    std::bind(&CountUntilServerNode::executeCallback, this, _1)
);
```

For actions, the C++ syntax is stricter than Python. On top of the action interface, object to link to, and action name, you have to provide three callbacks (even if you don't want to use them all):

- **Goal callback**: To accept or reject incoming goals.
- **Cancel callback**: To receive cancel requests.
- **Execute callback**: This is called the *handle accepted callback* in C++, but I named it *execute callback* to make the code similar to the Python one. In this callback, we execute goals that have been accepted.

> **Note**
>
> I've designed this chapter so that we write minimal code first, and then add the extra optional features. However, the C++ `create_server()` method will not work if you don't provide a cancel callback. Thus, what we will do for now is add this callback but not fully implement the cancel mechanism; we'll do that later.

At this point, we need to implement the three callback methods.

Implementing the callbacks

The arguments inside the callbacks can be quite long to write. That's why I suggest simplifying the code with `using` lines at the beginning, as well as double-checking everything as it's easy to make mistakes.

Here's the beginning of the goal callback method:

```cpp
rclcpp_action::GoalResponse goalCallback(const rclcpp_action::GoalUUID &uuid, std::shared_ptr<const CountUntil::Goal> goal)
```

Here, you get a unique identifier for the goal and the goal itself (to be precise, this is a `const` shared pointer to the goal). In the callback, we validate the goal and then accept or reject it. For example, to accept the goal, you would return the following:

```cpp
return rclcpp_action::GoalResponse::ACCEPT_AND_EXECUTE;
```

The next callback method is the cancel callback, in which you can decide whether to accept or reject an incoming cancel request. As I will explain the cancel mechanism later in this chapter, I will skip this part now—you just have to write the callback so that the code can compile.

The most important callback here is the execute callback. In this method, we receive a goal handle (`const std::shared_ptr<CountUntilGoalHandle> goal_handle`). The first thing we must do is extract the data from the goal and initialize a few things:

```
int target_number = goal_handle->get_goal()->target_number;
double delay = goal_handle->get_goal()->delay;
auto result = std::make_shared<CountUntil::Result>();
int counter = 0;
rclcpp::Rate loop_rate(1.0/delay);
```

You've probably started to get used to seeing shared pointers everywhere, and here is no exception. We don't create a result object, but a shared pointer to a result object.

Then, to handle the waiting time between each count iteration, we use a `rclcpp::Rate` object. This is a bit different from what we did with Python. In this rate object, we have to pass the rate—that is, the frequency we want for the loop. For example, if the delay is 0.5 seconds, the frequency would be 2.0 Hz. We can now execute the action:

```
RCLCPP_INFO(this->get_logger(), "Executing the goal");
for (int i = 0; i < target_number; i++) {
    counter++;
    RCLCPP_INFO(this->get_logger(), "%d", counter);
    loop_rate.sleep();
}
```

Here, we use the `sleep()` function of the rate object to pause the execution.

Finally, once the `for` loop ends, we can finish the execution:

```
result->reached_number = counter;
goal_handle->succeed(result);
```

In Python, we would set the goal's final state first, and then return the result. In C++, we don't return anything (note the `void` return type). We send the result at the same time as setting the goal state.

> **Note**
> Writing C++ code with actions starts to be quite complex, especially if you don't have much C++ experience. If you feel completely lost, maybe either continue with Python only or, as mentioned previously, skip this chapter for now and come back to it later.

That's it for the C++ action server. We can now write the client node and try the communication.

Writing an action client

We now have the minimal code required for the server to receive a goal, accept it, execute it, and return a result. At this point, we can write the client side of the communication.

The action client will send a goal to the server. It will then register a callback to find out whether the goal was accepted or rejected. If the goal is accepted, the client will register yet another callback to get the final result. That's what we're going to implement now—first with Python, then with C++.

Where should you write the action client? In your own ROS 2 applications, you could add an action client to any node. As an example, let's say you have a node that monitors the battery level of a mobile robot. This node could already have some publishers, subscribers, services, and so on. On top of all that, you can add an action client that will send a goal to another node (such as the server node, which controls the wheels of the robot) when the battery runs low.

For this chapter, and to keep things simple, we will create a new node, just for the action client. You can then use this code as a template for adding an action client anywhere you want. You can find the code for this section in `count_until_client_minimal` (`.py` or `.cpp`).

Let's start with the Python action client.

Writing a Python action client

Create a new Python file named `count_until_client_minimal.py` in the `my_py_pkg` package. Make this file executable.

Creating an action client

Let's start by setting up the action client. First, add the dependencies we will need:

```
import rclpy
from rclpy.node import Node
from rclpy.action import ActionClient
from rclpy.action.client import ClientGoalHandle, GoalStatus
from my_robot_interfaces.action import CountUntil
```

As for the action server, we don't get the action client directly from the `Node` class. Instead, we have to import `ActionClient` from `rclpy.action`.

We must also import the action interface, which should be the same as for the server. If we import this interface, we also need to add a dependency to the `package.xml` file. However, we have already done that, so there's no need to add anything else.

Then, in the node's constructor, we create an action client:

```
self.count_until_client_ = ActionClient(
self, CountUntil, "count_until")
```

We use the `ActionClient` class directly, and we pass three arguments: the object to bind to (`self`), the action interface, and the action name. Double-check that the name is the same as on the server side.

Then, to send a goal to the server, we add a new method:

```
def send_goal(self, target_number, delay):
    self.count_until_client_.wait_for_server()
    goal = CountUntil.Goal()
    goal.target_number = target_number
    goal.delay = delay
    self.count_until_client_.send_goal_async(
        goal).add_done_callback(self.goal_response_callback)
```

Here are the steps for sending a goal from the client to the server:

1. You can wait for the server with `wait_for_server()`. If you send a goal when the server isn't up and running, you will get an error, so ensure it's ready before you do anything. I didn't provide a timeout here, so it will wait indefinitely. You could add a timeout and do something similar to what we did in *Chapter 6*, in the *Writing a service client* section.
2. Create a goal object from the interface: `Interface.Goal()`.
3. Fill in the goal fields. Any field you omit will get a default value (0 for numbers, " " for strings).
4. Send the goal with `send_goal_async()`. This will return a Python `Future` object.
5. Register a callback for the goal's response so that you know it's been accepted or rejected.

Note that just as for services, we make an asynchronous call with `send_goal_async()`. This way, the method will return and we won't block the execution. If we were to block the execution, we would also block the spin, and thus we would never get any response.

Implementing the callbacks

So far, we've sent a goal with the action client and registered a callback, `goal_response_callback()`. Let's implement this method:

```
def goal_response_callback(self, future):
    self.goal_handle_: ClientGoalHandle = future.result()
    if self.goal_handle_.accepted:
        self.get_logger().info("Goal got accepted")
        self.goal_handle_.get_result_async().add_done_callback(
            self.goal_result_callback)
```

```
        else:
            self.get_logger().info("Goal got rejected")
```

In this callback, we get a `ClientGoalHandle` object from the result of the Python `Future` object. From this goal handle, we can find out whether the goal was accepted or not.

Please note that you won't get the final result in this goal response callback. Here, we only get to know whether the server accepted the goal or not. If the goal is accepted, we know that the server will start executing it and return a result at some point.

Then, in the client, we can register another callback for the goal result:

```
def goal_result_callback(self, future):
    status = future.result().status
    result = future.result().result
    if status == GoalStatus.STATUS_SUCCEEDED:
        self.get_logger().info("Success")
    elif status == GoalStatus.STATUS_ABORTED:
        self.get_logger().error("Aborted")
    elif status == GoalStatus.STATUS_CANCELED:
        self.get_logger().warn("Canceled")
    self.get_logger().info("Result: " + str(result.reached_number))
```

In this callback, we get the goal's final state and result after the server has finished executing the goal.

You can do anything you want with this result—here, we simply print it. As you can see, we will receive any of those three final states for the goal: STATUS_SUCCEEDED, STATUS_ABORTED, and STATUS_CANCELED.

Finally, let's not forget to call the `send_goal()` method. We will do this in the `main()` function, just after we initialize the node, and before we make the node spin:

```
node = CountUntilClientNode()
node.send_goal(5, 0.5)
rclpy.spin(node)
```

This will ask the server to count until 5 and wait 0.5 seconds between each count.

Trying the communication

We can now try the communication between the client and server.

Create an executable (in `setup.py`) for both the client and server nodes. Build the package and source the environment.

Then, start the server node and the client node in two different Terminals. You should see some logs in both Terminals as the communication progresses. In the end, you will get something like this for the server:

```
$ ros2 run my_py_pkg count_until_server
[count_until_server]: Action server has been started.
[count_until_server]: Received a goal
[count_until_server]: Accepting the goal
[count_until_server]: Executing the goal
[count_until_server]: 1
...
[count_until_server]: 5
```

For the client:

```
$ ros2 run my_py_pkg count_until_client
[count_until_client]: Goal got accepted
[count_until_client]: Success
[count_until_client]: Result: 5
```

You can see the flow of execution with the timestamp in each log. Here, we tested the case when the target number was positive—and thus, the goal was accepted. If you want, you can also test the case when the target number is negative; you should see the goal being rejected and not executed.

Now, let's learn how to write an action client with C++.

Writing a C++ action client

For the C++ code, I will focus on the few important points to notice in the `count_until_client_minimal.cpp` file.

First, we have all the includes and `using` lines. Those are almost the same as for the C++ action server. However, for the goal handle, we get `ClientGoalHandle` (this was `ServerGoalHandle` in the server code):

```
using CountUntilGoalHandle =
rclcpp_action::ClientGoalHandle<CountUntil>;
```

To create an action client, we declare the client as a private attribute of the class:

```
rclcpp_action::Client<CountUntil>::SharedPtr count_until_client_;
```

Then, we initialize the client in the constructor:

```
count_until_client_ = rclcpp_action::create_client<CountUntil>(this,
"count_until");
```

As you can see (but that shouldn't be a surprise anymore), we store a shared pointer to the action client. When initializing it, we provide the action interface, the object to bind to (`this`), and the action name, which should be the same as the one defined in the server code.

At this point, we can create a `sendGoal()` method to send a goal to the server. This method follows the same steps as for the Python client. We wait for the server, then create a goal, fill in the goal fields, send the goal, and register a callback. However, there is a big difference in how we handle the callbacks:

```cpp
auto options = rclcpp_action::Client<CountUntil>::SendGoalOptions();
options.goal_response_callback = std::bind(
    &CountUntilClientNode::goalResponseCallback, this, _1);
options.result_callback = std::bind(
    &CountUntilClientNode::goalResultCallback, this, _1);
count_until_client_->async_send_goal(goal, options);
```

In Python, we would chain the callbacks after sending the goal. In C++, you first need to create a `SendGoalOptions` object. In this object, you can register the different callback methods for your client. Here, we register the response and the result callback. Then, you must pass this object to the `async_send_goal()` method. This will register all the callbacks for when the node is spinning.

Now that we've registered two callbacks, we need to implement them.

In the goal response callback, to check if the goal was accepted or rejected, we can simply write the following:

```cpp
if (!goal_handle) {
```

If this returns `false`, we know the goal was rejected. If it returns `true`, there's no need to do anything else in this callback as the result callback was already registered with the `SendGoalOptions` object.

In the result callback, we get the goal's final state with `result.code`. We can then compare it with the different codes in `rclcpp_action::ResultCode`, which are SUCCEEDED, ABORTED, and CANCELED. To get access to the actual result, we write `result.result`. This will be a shared pointer to the result object.

Finally, let's not forget to call the `sendGoal()` method in the `main()` function:

```cpp
auto node = std::make_shared<CountUntilClientNode>();
node->sendGoal(5, 0.5);
rclcpp::spin(node);
```

That's about it for the C++ action client. After writing both the client and server, create an executable for both (in `CMakeLists.txt`); then, build, source, and run the two nodes. You can even try running the Python client with the C++ server, or any other combination.

Now that both the client and server are running correctly, we can add the extra functionalities we get with actions: feedback and cancel.

Taking advantage of all the action mechanisms

The reason I'm talking about feedback and cancel mechanisms now and didn't previously is to try not to overwhelm you with too much code at once. I know that actions are more complex than everything you've seen before with ROS 2. The minimal code alone is already quite long and contains lots of small details you must pay attention to.

Also, as explained in the first part of this chapter, the feedback and cancel mechanisms are optional. You can create a fully working client/server communication without them.

We're now going to improve the minimal code and add a few more functionalities so that we can take full advantage of ROS 2 actions. Here's what you can do to prepare the files for this section:

1. Make a copy of the files containing `_minimal`.
2. Rename those new files by removing the `_minimal`.

For example, you can make a copy of `count_until_client_minimal.py` (we won't modify this file anymore) and rename the copy `count_until_client.py` (this is where we will add more code). You can find the same organization in this book's GitHub repository.

So, let's explore the feedback and cancel mechanisms, starting with feedback, which is the easiest one.

Adding the feedback mechanism

When we wrote the action interface, we had to define three things: goal, result, and feedback. So far, we've only used the goal and result. The feedback is optional, and you could choose to leave it empty in the action definition. In this case, there's nothing else to do.

Since we've defined feedback in `CountUntil.action (int64 current_number)`, let's use it in our code so that we can make the server send feedback every time it increases the counter. The action client will be able to receive this feedback inside a callback.

Feedback with Python

Let's start with the action server. There are just a few lines to add so that we can publish the feedback.

Open `count_until_server.py`. In the `execute_callback()` method, at the same time as creating a result object, create a feedback object:

```
feedback = CountUntil.Feedback()
```

Now, when you execute the goal, you have to do the following:

```
feedback.current_number = counter
goal_handle.publish_feedback(feedback)
```

We must fill in the different fields of the feedback object and then send the feedback to the client with the `publish_feedback()` method from the goal handle.

That's all there is to it for the server side. Now, let's write the code to receive the feedback.

Open the `count_until_client.py` file and modify the line where you send the goal with `send_goal_async()`:

```
self.count_until_client_.send_goal_async(
    goal, feedback_callback=self.goal_feedback_callback). \
    add_done_callback(self.goal_response_callback)
```

To get the feedback with a Python action client, you must register a callback function when you send the goal. Here's the implementation for this callback:

```
def goal_feedback_callback(self, feedback_msg):
    number = feedback_msg.feedback.current_number
    self.get_logger().info("Got feedback: " + str(number))
```

With this, we get a feedback message and can access each field of that message. You can do anything you want with this feedback. For example, if your action client is asking for a robot to move to certain (x, y) coordinates, you might receive feedback on the current progress of the robot. From this, you could take any appropriate measure: cancel the goal (see the next section), send a new goal, and so on.

That's it regarding feedback. You can build your package again, source it, and run the two nodes. Here's what you will see on the client side:

```
$ ros2 run my_py_pkg count_until_client
[count_until_client]: Goal got accepted
[count_until_client]: Got feedback: 1
```

It will continue as follows:

```
[count_until_client]: Got feedback: 5
[count_until_client]: Success
[count_until_client]: Result: 5
```

With this feedback, the client isn't in the dark anymore. It can get to know what's happening between sending the goal and receiving the result.

Feedback with C++

The behavior for adding the feedback for the action server in `count_until_server.cpp` is the same as it is for Python.

First, you must create a feedback object in the execute callback:

```
auto result = std::make_shared<CountUntil::Result>();
```

The only difference is that we use a shared pointer here.

Then, you must publish the feedback:

```
feedback->current_number = counter;
goal_handle->publish_feedback(feedback);
```

On the client side, the way a callback is registered is a bit different. Open `count_until_client.cpp` and add the following line to the `sendGoal()` method:

```
options.feedback_callback = std::bind(
    &CountUntilClientNode::goalFeedbackCallback, this, _1, _2);
```

For a C++ action, we register all callbacks in the `SendGoalOptions` object that we pass to the `async_send_goal()` method.

Then, you can implement the callback:

```
void goalFeedbackCallback(const CountUntilGoalHandle::SharedPtr &goal_handle, const std::shared_ptr<const CountUntil::Feedback> feedback)
{
    (void)goal_handle;
    int number = feedback->current_number;
    RCLCPP_INFO(this->get_logger(), "Got feedback: %d", number);
}
```

Here, we receive both the goal handle and the feedback (as `const` shared pointers).

> **Note**
>
> As you can see, whenever there's an argument we don't use in a function, I write `(void)`, followed by the argument. This is a way to prevent getting *unused parameter* warnings when compiling with `colcon build`. As a best practice, you should address all errors and warnings in your code when developing a ROS 2 application. If you don't do this, you will end up with lots of ignored warnings, and you could miss the important ones, leading to hard-to-debug issues in the future.

Now that the code is complete, you can compile the package and run the client and server nodes in two different Terminals. You should see a similar output to what we had for Python.

Implementing the feedback mechanism is relatively easy. Now, let's learn how to cancel a goal. This will be more complex and require the use of more advanced ROS 2 concepts.

Adding the cancel mechanism

After sending a goal, the client can decide to ask the server to cancel it. The server will receive this request and accept (or not) to cancel the goal. If the cancel request is accepted, the server will take any appropriate action to cancel the execution of the goal. In the end, the server will still send a result to the client.

What do we need to do in the code? In the server node, we will add another callback so that we can receive cancel requests and decide to accept or reject them. Then, in the execute callback, we will be able to check whether the goal should be canceled; if so, we will terminate the execution sooner.

However, if we just do this, it's not going to work and the cancel requests will never be received. Why is that? Let's explore this question now.

> **Note**
> This section introduces a few concepts that are outside the scope of this (beginner) book. I will talk about them briefly without going into full detail. If you'd like to understand these in more depth, feel free to explore the advanced concepts by yourself (you will find additional resources in *Chapter 14*). You can see this section as a going further with actions section.

Understanding the problem with cancel and spin

We will only focus on the server side here as this is where the issue will occur. I will explain what the issue is so that we can implement the solution later.

So, when you start the action server, three callbacks will be registered: a goal callback, a cancel callback, and an execute callback.

With our current code, when the server receives a goal, here's what happens:

1. The goal is received by the goal callback and is accepted or rejected.
2. If accepted, we execute the goal in the execute callback. Something crucial to note is that while we execute the goal with the `for` loop, the thread is blocked.
3. Once the goal is executed, we return the result and exit from the execute callback.

The problem is with *Step 2*. Since we're blocking the execution, we're blocking the spin mechanism.

When you make a node spin, what's happening? As mentioned previously, the node will be kept alive and all callbacks can be processed. However, the spin is working in a single thread. This means that if you have one callback taking 5 seconds to execute, it will block the following callbacks for 5 seconds.

We never had any issues before because all the callbacks we wrote were very quick to execute. However, with the execute callback for an action, we're in a situation where the execution could take quite some time, and thus block all the other callbacks.

That's quite the problem. How can you ask to cancel a goal if the cancel request is only received after the goal's execution has finished?

To solve this problem, we have two possible solutions:

- **The classic programming way**: We could create a new thread in the execute callback. The callback can then exit while the goal is processed in the background. The spin continues, and thus, other callbacks can be called.
- **The ROS 2 way**: We can use a multi-threaded executor, which means that our spin mechanism will work not in a single thread, but in multiple threads. Thus, if one callback is blocking, you can still execute other callbacks—including the cancel callback.

Since we want to follow ROS 2 principles to stay consistent with other developers, we're going to follow the ROS 2 way and solve that issue with a multi-threaded executor.

> **Note**
> I'm not going to go into more detail about single and multi-threaded executors here. I'm using them now so that we can implement the cancel mechanism correctly. Executors can be a great topic to explore after reading this book.

The process for the cancel mechanism in the server code will be the same for Python and C++:

1. Register a callback to handle cancel requests.
2. Cancel the goal in the execute callback.
3. Make the node spin with a multi-threaded executor.

Canceling with Python

We will start with the server code, which can be found in `count_until_server.py`.

First, let's register a callback to receive cancel requests:

```
ActionServer(
    …
    cancel_callback=self.cancel_callback,
    …)
```

Here's the callback's implementation:

```
def cancel_callback(self, goal_handle: ServerGoalHandle):
    self.get_logger().info("Received a cancel request")
    return CancelResponse.ACCEPT
```

In this callback, you receive a goal handle corresponding to the goal the client wants to cancel. You can then create any kind of condition to decide whether the goal should be canceled or not. To accept, you must return `CancelResponse.ACCEPT`; to reject, you must return `CancelResponse.REJECT`. With this example, I kept things simple and we just accepted the cancel request without implementing any other checks.

Now, if the cancel request has been accepted, we need to do something about it. In the execute callback, while we're executing the goal (inside the `for` loop), add the following code:

```
if goal_handle.is_cancel_requested:
    self.get_logger().info("Canceling goal")
    goal_handle.canceled()
    result.reached_number = counter
    return result
```

When we accept a cancel request, an `is_cancel_requested` flag in the goal handle will be set to `True`. Now, in the execute callback, we simply need to check this flag.

What we do in the code is stop the current execution. If, for example, your action server controls the wheels of a robot, you could interpret `cancel` as "decelerate and stop moving," "step on the side so we don't block the main way," or even "go back to base." The way you handle the behavior for the cancellation depends on each application. Here, we just stop counting.

In the execute callback, you need to set the goal's final state and return a result, even if you cancel the goal. Thus, we use the `canceled()` method to set the state, and we return a result that contains the last reached number. If the client asks the server to count to 10 and then cancels the goal when the counter is at 7, the result will contain 7.

That's it for the cancel mechanism. However, to make things work, as we've seen previously, we need to use a multi-threaded executor.

First, you'll need to import the following:

```
from rclpy.executors import MultiThreadedExecutor
from rclpy.callback_groups import ReentrantCallbackGroup
```

When using multi-threaded executors, we also need to use *callback groups*. Here, `ReentrantCallbackGroup` will allow all callbacks to be executed in parallel. This means that you can have several goal, cancel, and execute callbacks running at the same time for one action server.

When you create the action server, add a `callback_group` argument:

```
ActionServer(
    ...
    callback_group=ReentrantCallbackGroup())
```

Finally, modify the line to make the node spin in the `main()` function:

```
rclpy.spin(node, MultiThreadedExecutor())
```

That's all there is to it. It's just a few lines of code, but adding this requires a good understanding of ROS 2 and its underlying mechanisms.

Let's write the code for the client so that we can send a cancel request for a goal that's being executed. In `count_until_client.py`, add a method to cancel a goal:

```
def cancel_goal(self):
    self.get_logger().info("Send a cancel goal request")
    self.goal_handle_.cancel_goal_async()
```

Here, we're using the goal handle that we saved in the goal response callback (`self.goal_handle_: ClientGoalHandle = future.result()`). From this goal handle object, we have access to a `cancel_goal_async()` method.

So, where do we cancel the goal? This can be done from anywhere: from the feedback callback, an independent subscriber callback, and so on. It will depend on your application.

To make a quick test, let's arbitrarily decide that we want to cancel the goal if the `current_number` field from the feedback is greater than or equal to 2. It doesn't make any sense (why would we ask to count until 5, only to cancel if the number reaches 2?), but it's a quick way to test the cancel mechanism.

In the goal feedback callback, add the following code:

```
if number >= 2:
    self.cancel_goal()
```

Then, build the package, source it, and run both the server and client. Here's the log for the client:

```
[count_until_client]: Goal got accepted
[count_until_client]: Got feedback: 1
[count_until_client]: Got feedback: 2
[count_until_client]: Send a cancel goal request
[count_until_client]: Canceled
[count_until_client]: Result: 2
```

For the server, you will see this:

```
[count_until_server]: Executing the goal
[count_until_server]: 1
[count_until_server]: 2
[count_until_server]: Received a cancel request
[count_until_server]: Canceling goal
```

With this, we used all the mechanisms available for actions. Now, you can comment the lines to cancel the goal from the feedback callback—this was just for testing purposes.

Canceling with C++

In the server code (`count_until_server.cpp`), we added a cancel callback when we created the action server. This was mandatory so that the code could compile.

In this callback, we just accept the cancel request:

```
return rclcpp_action::CancelResponse::ACCEPT;
```

Then, to handle the cancellation of the goal in the execute callback, add the following code to the `for` loop:

```
if (goal_handle->is_canceling()) {
    RCLCPP_INFO(this->get_logger(), "Canceling goal");
    result->reached_number = counter;
    goal_handle->canceled(result);
    return;
}
```

In C++, we check the `is_canceling()` method inside the goal handle. If it returns `true`, this means that a cancel request for this goal has been accepted, and we need to do something about it.

We set the goal's final state and result with `canceled()`, and we exit from the execute callback.

That's it for the cancel mechanism, but now we need to make the node spin with a multi-threaded executor.

In the `main()` function, we must replace the `rclcpp::spin(node);` line with the following code:

```
rclcpp::executors::MultiThreadedExecutor executor;
executor.add_node(node);
executor.spin();
```

Here, we create an executor, add the node, and make the executor spin. Then, as we did for Python, inside the node, we need to add a callback group. We can declare one as a private attribute:

```
rclcpp::CallbackGroup::SharedPtr cb_group_;
```

Finally, we modify the code in the node's constructor to give a reentrant callback group to the action server, so that all callbacks can be executed in parallel:

```
cb_group_ = this->create_callback_group(
    rclcpp::CallbackGroupType::Reentrant);
count_until_server_ = rclcpp_action::create_server<CountUntil>(
    ...
```

```
        rcl_action_server_get_default_options(),
        cb_group_
);
```

We also need to add `rcl_action_server_get_default_options()` after the callbacks and before the callback group; otherwise, the compiler will complain about not finding an overload for the `create_server()` function.

Now that we've finished writing the server code, let's send a cancel request from the client. In `count_until_client.cpp`, add a `cancelGoal()` method:

```
void cancelGoal()
{
    RCLCPP_INFO(this->get_logger(), "Send a cancel goal request");
    count_until_client_->async_cancel_all_goals();
}
```

In C++, we cancel goals from the action client, not from the goal handle. To make things simpler here, we're canceling all goals that could have been sent by this client.

To test the cancel mechanism, we add those lines to the feedback callback:

```
if (number >= 2) {
    cancelGoal();
}
```

Once you've completed the code, run your C++ action client and server nodes. You can also try any combination of Python and C++ nodes; they should behave the same way. Once you've tested your code, comment the lines to cancel the goal from the feedback callback.

Let's finish this chapter with a few more command-line tools that will help you when you're developing applications with actions.

Additional tools to handle actions

Since actions are part of the core ROS 2 functionalities, they also get their own command-line tool: `ros2 action`.

In this section, we'll learn how to introspect actions, send a goal from the Terminal, and change an action name at runtime.

To see all the possible commands, type `ros2 action -h`.

Listing and introspecting actions

Actions are based on topics and services. Since `rqt_graph` doesn't support services (for now), we could see the topics for an action server and client, but that's about it. Thus, `rqt_graph` won't be very useful for introspecting actions. Because of this, we will use the `ros2` command-line tool here.

Let's learn how to find existing actions and how to get the interface for one action.

Stop all nodes and start the `count_until_server` node (Python or C++ one). Then, list all available actions by running the following command:

```
$ ros2 action list
/count_until
```

Here, we found the `/count_until` action. As we've seen with topics and services, if you don't provide any namespace for the name (we wrote `count_until` in the server code), a leading slash will be added automatically.

From this action name, we can get more information, including the action interface.

Run `ros2 action info <action_name> -t`:

```
$ ros2 action info /count_until -t
Action: /count_until
Action clients: 0
Action servers: 1
/count_until_server [my_robot_interfaces/action/CountUntil]
```

From this, we can see that the action server is hosted in the `count_until_server` node, and we also find the action interface. For `ros2 action info` to show the interface, don't forget to add `-t`; otherwise, you'll just see the node's name.

Finally, we can get the interface:

```
$ ros2 interface show my_robot_interfaces/action/CountUntil
# Here you should see the action definition
```

This process is the same as what we followed for services. Now that we know the action name and interface, we can try the service directly from the Terminal.

Sending a goal from the Terminal

If you write a service server and want to try it before writing the action client, you can use the `ros2 action send_goal` command line.

The complete command is `ros2 action send_goal <action_name> <action_interface> "<goal_in_json>"`. You can also add `--feedback` after the command to receive the (optional) feedback from the server. Let's try it out:

```
$ ros2 action send_goal /count_until my_robot_interfaces/action/
CountUntil "{target_number: 3, delay: 0.4}" --feedback
```

You will get the following result:

```
Waiting for an action server to become available...
Sending goal:
 target_number: 3
delay: 0.4
Goal accepted with ID: cad1aa41829d42c5bb1bf73dd4d66600
Feedback:
current_number: 1
Feedback:
current_number: 2
Feedback:
current_number: 3
Result:
reached_number: 3
Goal finished with status: SUCCEEDED
```

This command is very useful for developing action servers. However, it will only work well for actions for which the goal is simple. Here, we only have an integer number and a double number. If the goal contains an array of 20 3D points, you will spend more time trying to write the command correctly than implementing an action client. In this case, to go faster, use the action client we've written in this chapter as a template.

Topics and services inside actions

By default, with `ros2 topic list` and `ros2 service list`, you won't see the two topics and three services inside the action. However, they do exist—you just have to add `--include-hidden-topics` and `--include-hidden-services`, respectively:

```
$ ros2 topic list --include-hidden-topics
/count_until/_action/feedback
/count_until/_action/status
...
$ ros2 service list --include-hidden-services
/count_until/_action/cancel_goal
/count_until/_action/get_result
/count_until/_action/send_goal
...
```

With that, we've found all the topics and services that are being used. You can explore these a bit more by yourself by using the other `ros2 topic` and `ros2 service` command lines.

Now, there's one thing we've done for nodes, topics, and services: we've changed the name at runtime. For some reason, this feature isn't available for actions yet. As a workaround, you can still do this by renaming the two topics and three services when you start the action server:

```
$ ros2 run my_cpp_pkg count_until_server --ros-args \
    -r /count_until/_action/feedback:=/count_until1/_action/feedback \
    -r /count_until/_action/status:=/count_until1/_action/status \
    -r /count_until/_action/cancel_goal:=/count_until1/_action/cancel_goal \
    -r /count_until/_action/get_result:=/count_until1/_action/get_result \
    -r /count_until/_action/send_goal:=/count_until1/_action/send_goal
```

With this, the action will be renamed /count_until1. The command is a bit ugly and prone to errors, but when we start nodes using launch files in *Chapter 9*, in the *Configuring nodes inside a launch file* section, this won't be a problem.

With that, we've come to the end of this chapter. I haven't added any challenges here as I think that this chapter itself is a big enough challenge. I would prefer you to spend your time continuing with the other concepts in this book instead of being stuck too long on actions, especially if you're just getting started with ROS.

Summary

In this chapter, you worked on ROS 2 actions. You created various actions to solve a problem that services don't handle well: when the server may take some time to execute the request.

With actions, you can properly handle this case. While the goal is being executed, you can get some feedback from the server, or even decide to cancel the goal. Also, you could handle several goals at the same time, queue them, replace one with another one, and so on (we haven't seen this in this chapter as it's something you can look into if you want to develop your skills further).

You can implement action servers and clients in your code using the `rclpy.action` library for Python and the `rclcpp_action` library for C++.

Here are the main steps for writing an action server:

1. Since we're on the server side, we must choose the action name and interface. Usually, for an action, you will have to create a custom interface (in a dedicated package).

2. Then, you must import the interface into your code and create an action server in the constructor. Here, you will register three callback methods:

- **Goal callback**: When the server receives a goal, choose whether to accept or reject it.
- **Execute callback**: After a goal has been accepted, you can execute it. During the execution of the goal, you can also publish optional feedback.
- **Cancel callback (optional mechanism)**: If you receive a cancel request, you can accept or reject it. If accepted, you will have to cancel the current goal execution.

To write an action client, you must follow these steps:

1. Find which name and interface you need to use so that you can communicate with the server.
2. Import the interface into your code and create an action client in the constructor.
3. Add a method to send a goal. After you send a goal, you will have to write several callbacks:

   - **Goal response callback**: You will know whether the goal has been accepted or rejected by the server.
   - **Goal result callback**: After the goal has been executed by the server, you will get the result and the goal's final state here.
   - **Feedback callback (optional)**: If the server publishes any feedback, you can receive it here.

4. Finally, from anywhere in the code, you can decide to cancel the execution of a currently active goal.

On top of all that, with the `ros2 action` command line, you can introspect your actions and send goals directly from the Terminal. Also, since actions are based on topics and services, you can introspect each underlying communication with `ros2 topic` and `ros2 service`, respectively.

Now, if you managed to get here while reading this book for the first time, congratulations—this chapter is probably one of the most difficult to follow. If you're still wondering what I was talking about the whole time, don't worry—you can come back to actions later once you've finished this book and become more experienced with ROS.

We're now done with the three types of communication in ROS 2. In the next chapter, we will go back to a more beginner level and continue to work on nodes. This time, we will learn how to customize nodes when we start them so that we can make our application more dynamic.

8
Parameters – Making Nodes More Dynamic

We are now done with the basics of ROS 2 communications. In this chapter, we will continue to work on nodes, but this time by making them more dynamic with **parameters**.

To understand parameters, I will start with why we need them in the first place. Then, you will learn how to add parameters to your nodes so that you can customize them at runtime. You will also see how to load multiple parameters at once with **YAML** files and how to allow parameters to be modified in your code with **parameter callbacks**.

As a starting point, we will use the code inside the `ch7` folder of the book's GitHub repository (`https://github.com/PacktPublishing/ROS-2-from-Scratch`). If you skipped *actions* (*Chapter 7*), you can also start from the `ch6` folder, which will work the same. The final code for this chapter will be in the `ch8` folder.

By the end of this chapter, you will be able to add parameters to any of your nodes and handle parameters for other nodes that you start.

The concept of parameters is not too difficult, and there won't be too much to do in the code. However, it's an important concept and the first step toward making your application more dynamic and scalable.

In this chapter, we will cover the following topics:

- What is a ROS 2 parameter?
- Using parameters in your nodes
- Storing parameters in YAML files
- Additional tools to handle parameters
- Updating parameters with parameter callbacks
- Parameter challenge

What is a ROS 2 parameter?

You have already experimented a bit with parameters in *Chapter 3*, where you ran a node with different settings.

I am now going to start from scratch again and explain parameters with a real-life example.

Why parameters?

Let's start with a problem to understand the need for parameters. I will use a camera driver as an example—we won't write the node; it's just for the explanation.

This camera driver connects to a USB camera, reads images, and publishes them on a ROS 2 topic. This is classic behavior for any ROS 2 hardware driver.

Inside this node, you will have some variables for different kinds of settings. Here are some examples:

- USB device name
- Frames per second (FPS)
- Simulation mode

Let's say the camera you're working on is connected to the `/dev/ttyUSB0` port (typical USB port name on Linux). You want to set `60` FPS and not use the simulation mode (`false`). Those are the values you will write for the variables inside your node.

Later on, if the USB device name is different (for example, `/dev/ttyUSB1`), you will have to change that setting in your code and maybe build again—you'll do the same thing if you want to start your camera with `30` FPS instead of `60` FPS, or if you want to run it in simulation mode.

Also, what if you have two cameras, and you want to use them both at the same time? Will you duplicate your code for each camera? How can you handle the different settings for both cameras?

As you can see, hardcoding those settings in your code is not a great option for reusability. This is why we have ROS 2 parameters.

Example of a node with parameters

A ROS 2 parameter is basically a setting for a node that you can modify when you start the node.

So, if we keep the camera driver example, we could add three parameters—USB device name (string), FPS value (integer), and simulation mode (boolean):

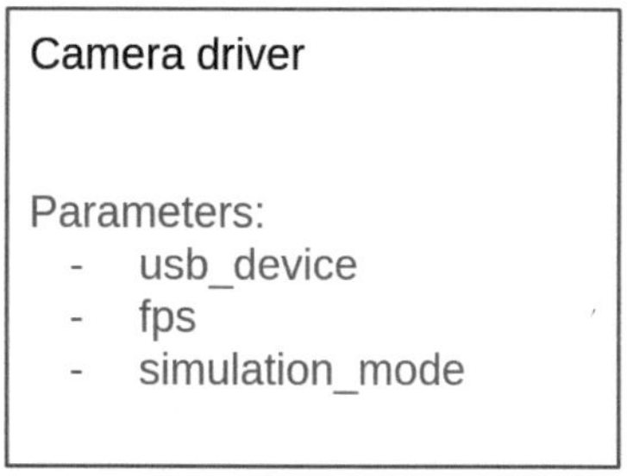

What is a ROS 2 parameter?   189

```
┌─────────────────────────────┐
│ Camera driver               │
│                             │
│ Parameters:                 │
│   -  usb_device             │
│   -  fps                    │
│   -  simulation_mode        │
└─────────────────────────────┘
```

Figure 8.1 – A node class with three parameters

When you start this camera driver with `ros2 run` (we will see how to do that later in this chapter), you will be able to provide the values you want for those three parameters.

Let's say you want to start two nodes for two different cameras, given the following settings:

1. Port: `/dev/ttyUSB0`; FPS: `30`; simulation mode: off
2. Port: `/dev/ttyUSB1`; FPS: `60`; simulation mode: off

With the parameters we've added in the code, we can start the same node multiple times with different values:

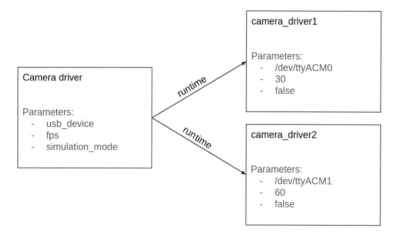

Figure 8.2 – Starting two nodes with different settings

From the same code, we start two different nodes. At runtime, we rename the nodes (because we can't have two nodes with the same name), and we provide the parameters' values.

Our two camera nodes are now running, each with a different configuration. You could stop one camera node and start it again with a different set of values.

ROS 2 parameters – wrapping things up

With parameters, you can reuse the same code and start several nodes with different settings. There's no need to compile or build anything again; you just have to provide the parameters' values at runtime.

Making your nodes customizable allows for greater flexibility and reusability. Your application will become much more dynamic.

Parameters are also very convenient for collaborating with other ROS developers. If you develop a node that could be reused by others, then with parameters, you allow other developers to fully customize the node without even having to look at the code. This also applies when using existing nodes. Lots of them can be configured at runtime.

Here are a few important points about parameters:

- Just as with a variable, a parameter has a name and a data type. Among the most common types, you can use booleans, integer numbers, float numbers, strings, and lists of those types.
- A parameter's value is specific to a node. If you kill the node, the value is gone with it.
- You can set the value for each parameter when you start a node with `ros2 run` (or from a launch file, which we will see in the next chapter).

Now, how to add parameters to your code? As for nodes, topics, and services, you will get everything you need from the `rclpy` and `rclcpp` libraries. You will be able to declare the parameters in your code and get the value for each parameter.

Using parameters in your nodes

We will continue with the code we have written in the previous chapters. Here, we will improve the `number_publisher` node. As a quick recap, this node publishes a number on a topic, at a given rate. The number and publishing rate are directly written in the code.

Now, instead of hardcoding the number and publishing rate values, we will use parameters. This way, we will be able to specify what number to publish, and the publishing frequency or period, when we start the node.

You need to follow two steps to be able to use a parameter in your code:

1. Declare the parameter in the node. This will make the parameter exist within the node so that you can set a value to it when starting the node with `ros2 run`.
2. Retrieve the parameter's value so that you can use it in the code.

Let's start with Python, and later on, we will also see the C++ code.

Declaring, getting, and using parameters with Python

Before using a parameter, we need to declare it. Where should we declare parameters? We will do that in the node's constructor, before everything else. To declare a parameter, use the declare_parameter() method from the Node class.

You will provide two arguments:

- **Parameter name**: This is the name that you will use to set the parameter's value at runtime
- **Default value**: If the parameter's value is not provided at runtime, this value will be used

There are, in fact, different ways to declare a parameter. You don't necessarily need to provide a default value if you provide the parameter type instead. However, we will keep things like that, as it will probably make your life easier. Adding a default value for each parameter is a best practice to follow.

Open the number_publisher.py file, and let's declare two parameters in the constructor:

```
self.declare_parameter("number", 2)
self.declare_parameter("publish_period", 1.0)
```

A parameter is defined by a name and a data type. Here, you choose the name, and the data type will be automatically set depending on the default value you have provided. In this example, the default value for number is 2, which means that the parameter's data type is integer. For the publish_period parameter, the default value is 1.0, which is a float number.

Here are a few more examples of different data types:

- **Booleans**: self.declare_parameter("simulation_mode", False)
- **String**: self.declare_parameter("device_name", "/dev/ttyUSB0")
- **Integer array**: self.declare_parameter("numbers", [4, 5, 6])

Now, declaring a parameter means that it exists within the node, and you can set a value from the outside. However, in your code, to be able to use the parameter, it's not enough to declare it. After doing that, you need to get the value.

For this, you will use the get_parameter() method, and provide the parameter's name as an argument. Then, you can access the value with the value attribute:

```
self.number_ = self.get_parameter("number").value
self.timer_period_ = self.get_parameter(
    "publish_period"
).value
```

At this point in the code, the number_ variable (which is a class attribute) contains the value that was set for the number parameter at runtime with ros2 run.

> **Note**
> You always need to declare a parameter before getting its value. If you fail to do so, when starting the node, you will get an exception (`ParameterNotDeclaredException`) as soon as you try to get the value.

After you get the values for all parameters and store them inside variables or class attributes, you can use them in your code. Here, we modify the timer callback:

```
self.number_timer_ = self.create_timer(
    self.timer_period_, self.publish_number
)
```

With this, we set the publishing period from the parameter's value.

That's pretty much it for the code. As you can see, there is nothing too complicated. For one parameter, you will just add two instructions: one to declare the parameter (give it a name and a default value), and another to get its value.

Now, I've been talking about setting a parameter's value at runtime with `ros2 run`. How do we do that?

Providing parameters at runtime

Before going further, make sure to save the `number_publisher.py` file and build the `my_py_pkg` package (if you haven't used `--symlink-install` before).

To provide a parameter's value with the `ros2 run` command, follow the next steps:

1. You will first start your node with `ros2 run <package_name> <exec_name>`.
2. Then, to add any argument after this command, you have to write `--ros-args` (only once).
3. To specify a parameter's value, write `-p <param_name>:=<param_value>`. You can add as many parameters as you want.

Let's say we want to start the node and publish the number 3 every `0.5` seconds. In that case, we'd run the following command:

```
$ ros2 run my_py_pkg number_publisher --ros-args -p number:=3 -p publish_period:=0.5
```

To verify it's working, we can subscribe to the `/number` topic:

```
$ ros2 topic echo /number
data: 3
---
data: 3
---
```

We can also verify the publish rate:

```
$ ros2 topic hz /number
average rate: 2.000
    min: 0.500s max: 0.500s std dev: 0.00004s window: 3
```

So, what happened? You provided some values for different parameters at runtime. The node will start and recognize those parameters because they match the names that have been declared in the code. Then, the node can get the value for each parameter.

If you provide the wrong data type for a parameter, you will get an error. As seen previously, the data type is set in the code from the default value. In this example, the `number` parameter should be an integer number. Look at what happens if we try to set a double value:

```
$ ros2 run my_py_pkg number_publisher --ros-args -p number:=3.14
…
    raise InvalidParameterTypeException(
rclpy.exceptions.InvalidParameterTypeException: Trying to set
parameter 'number' to '3.14' of type 'DOUBLE', expecting type
'INTEGER': number
[ros2run]: Process exited with failure 1
```

As you can see, once a parameter type is set in the code, you have to use that exact same type whenever you provide a value at runtime.

As each parameter has a default value, you could also omit one or more parameters:

```
$ ros2 run my_py_pkg number_publisher --ros-args -p number:=3
```

In this case, the `publish_period` parameter will be set to its default value (`1.0`), defined in the code.

To finish here, let's just see an example where renaming the node and setting parameters' values can allow you to run several different nodes from the same code without having to modify anything in the code.

In Terminal 1, run the following:

```
$ ros2 run my_py_pkg number_publisher --ros-args -r __node:=num_pub1
-p number:=3 -p publish_period:=0.5
```

In Terminal 2, run the following:

```
$ ros2 run my_py_pkg number_publisher --ros-args -r __node:=num_pub2 -p number:=4 -p publish_period:=1.0
```

With this, you have two nodes (num_pub1 and num_pub2), both publishing to the /number topic but with different data and publishing rates. With this example, you can see that parameters are a great way to make your nodes more dynamic.

Let's now finish this section with the C++ code for parameters.

Parameters with C++

Parameters work the same for Python and C++; only the syntax differs. Here, we will modify the number_publisher.cpp file.

In the constructor, you can declare some parameters:

```
this->declare_parameter("number", 2);
this->declare_parameter("publish_period", 1.0);
```

We use the declare_parameter() method from the rclcpp::Node class. The arguments are the same as for Python: name and default value. From this value, the parameter type will be set.

Then, to get a parameter's value in the code, write the following:

```
number_ = this->get_parameter("number").as_int();
double timer_period = this->get_parameter("publish_period")
                           .as_double();
```

We use the get_parameter() method and provide the name for the parameter. Then, we get the value with the method that corresponds to the data type: as_int(), as_double(), as_string(), as_string_array(), and so on. If you have an IDE with auto-completion, you should be able to see all possible types.

The rest is the same as for Python. Please refer to the GitHub files for any other minor changes and additions.

To start a node with parameters, run the following:

```
$ ros2 run my_cpp_pkg number_publisher --ros-args -p number:=4 -p publish_period:=1.2
```

Working with parameters is not that tough. For each parameter you want to create, you have to declare it in the code and get its value. When starting the node from the terminal, you can specify a value for each parameter you want.

Now, that works well only if you have a small number of parameters. In a real application, it's not uncommon to have a few dozen or even hundreds of parameters for a node. How can you manage so many parameters?

Storing parameters in YAML files

As your ROS 2 application grows, so will the number of parameters. Adding 10 or more parameters from the command line is not really an option anymore.

Fortunately, you can use YAML files to store your parameters, and you can load these files at runtime. If you don't know YAML, it's basically a markup language, similar to XML and JSON, but supposedly more readable by humans.

In this section, you will learn how to add your parameters to a YAML file and how to load this file at runtime.

Loading parameters from a YAML file

Let's start by saving parameters into a file so we can use them when we start a node.

First, create a YAML file with the `.yaml` extension. The filename doesn't matter that much, but it's better to give it a meaningful name. As our application deals with numbers, we can name it `number_params.yaml`.

For now, let's just create a new file in our home directory (in the next chapter, we will see how to properly install a YAML file in a ROS 2 application):

```
$ cd ~
$ touch number_params.yaml
```

Edit this file and add parameters for the `/number_publisher` node:

```
/number_publisher:
  ros__parameters:
    number: 7
    publish_period: 0.8
```

First, you write the name of the node. On the next line, and with an indentation (usually, it's recommended to use two spaces), we add `ros__parameters` (make sure you use two underscores). This will be the same for every node you add in a YAML file. On the following lines, and with yet another indentation, you can add all the parameters' values for the node.

> **Note**
> It's important that the node name matches; otherwise, the parameters won't be loaded into the node. If you omit the leading slash, it would still work for loading parameters with `ros2 run`, but you could have issues with other commands.

Once you've written this file, you can load the parameters with the `--params-file` argument:

```
$ ros2 run my_py_pkg number_publisher --ros-args --params-file ~/number_params.yaml
```

This will start the node and specify the values for the `number` and `publish_period` parameters.

If you have two or fifty parameters, the `ros2 run` command stays the same. All you have to do is add more parameters in the YAML file. If you want to modify a parameter, you can modify the corresponding line in the file or even create several files for different sets of configurations.

Parameters for multiple nodes

What should you do if you want to save parameters for several nodes?

Good news: inside one param YAML file, you can add the configuration for as many nodes as you want. Here's an example:

```yaml
/num_pub1:
  ros__parameters:
    number: 3
    publish_period: 0.5

/num_pub2:
  ros__parameters:
    number: 4
    publish_period: 1.0
```

This corresponds to the example we ran before, with two nodes and different parameters.

Now, to start the same nodes and parameters, we only need to run the commands shown next.

In Terminal 1, run the following:

```
$ ros2 run my_py_pkg number_publisher --ros-args -r __node:=num_pub1 --params-file ~/number_params.yaml
```

In Terminal 2, run the following:

```
$ ros2 run my_py_pkg number_publisher --ros-args -r __node:=num_pub2 --params-file ~/number_params.yaml
```

We give the same YAML file to both nodes. Each node will only load the parameters' values that are defined under the node name.

Recapping all parameters' data types

Let's say you have all those parameters declared in your code (Python example only, but you can easily translate to C++):

```
self.declare_parameter("bool_value", False)
self.declare_parameter("int_number", 1)
self.declare_parameter("float_number", 0.0)
self.declare_parameter("str_text", "Hola")
self.declare_parameter("int_array", [1, 2, 3])
self.declare_parameter("float_array", [3.14, 1.2])
self.declare_parameter("str_array", ["default", "values"])
self.declare_parameter("bytes_array", [0x03, 0xA1])
```

Those are basically all the available data types for parameters.

To specify the value for each parameter, you can create a YAML file or add some configuration to an existing YAML file. Here is what you would write for this node (named `your_node`):

```
/your_node:
  ros__parameters:
    bool_value: True
    int_number: 5
    float_number: 3.14
    str_text: "Hello"
    bool_array: [True, False, True]
    int_array: [10, 11, 12, 13]
    float_array: [7.5, 400.4]
    str_array: ['Nice', 'more', 'params']
    bytes_array: [0x01, 0xF1, 0xA2]
```

With YAML files, you will be able to customize your nodes in a quick and efficient way. I recommend using them as soon as you get more than a few parameters.

Also, as you continue your journey with ROS 2, you will start to use nodes and complete stacks developed by other developers. Those nodes often come with a bunch of YAML files that allow you to configure the stack without having to change anything in the nodes directly.

Let's now continue with the command-line tools. You have set the parameters' values with `ros2 run`, but there are actually more tools to handle parameters.

Additional tools for handling parameters

You start to get used to it: for each ROS 2 core concept, we get a dedicated `ros2` command-line tool. For parameters, we have `ros2 param`.

You can see all the commands with `ros2 param -h`. Let's focus on the most important ones so that we can get parameters' values from the terminal and set some values after the node has been started. At the end of this section, we will also explore the different parameter services available for all nodes.

Getting parameters' values from the terminal

After you've started one or several nodes, you can list all available parameters with `ros2 param list`.

Stop all nodes and start two nodes, num_pub1 and num_pub2, either by using the YAML file or by providing the parameters' values manually.

In Terminal 1, run the following:

```
$ ros2 run my_py_pkg number_publisher --ros-args -r __node:=num_pub1 -p number:=3 -p publish_period:=0.5
```

In Terminal 2, run the following:

```
$ ros2 run my_py_pkg number_publisher --ros-args -r __node:=num_pub2 -p number:=4 -p publish_period:=1.0
```

Now, list all available parameters:

```
$ ros2 param list
/num_pub1:
  number
  publish_period
  start_type_description_service
  use_sim_time
/num_pub2:
  number
  publish_period
  start_type_description_service
  use_sim_time
```

Here, I started two nodes to show you that each node gets its own set of parameters. The `number` parameter inside `/num_pub1` is not the same as the `number` parameter inside `/num_pub2`.

> **Note**
>
> For each parameter, we also always get the `use_sim_time` parameter with a default value of `false`. This means that we use the system clock. We would set it to `true` if we were simulating the robot so that we could use the simulation engine clock instead. This is not important for now, and you can ignore this parameter. You can also ignore the `start_type_description_service` parameter.

From this, you can get the value for one specific parameter, using `ros2 param get <node_name> <param_name>`:

```
$ ros2 param get /num_pub1 number
Integer value is: 3
$ ros2 param get /num_pub2 number
Integer value is: 4
```

This corresponds to the values we have set when starting the node. Using `ros2 param get` allows you to introspect the parameters inside any running node.

Exporting parameters into YAML

If you'd like to get the complete set of parameters for a node, you can do so with `ros2 param dump <node_name>`.

Let's dump all parameters for the nodes we are running.

For the first node, run the following:

```
$ ros2 param dump /num_pub1
/num_pub1:
  ros__parameters:
    number: 3
    publish_period: 0.5
    start_type_description_service: true
    use_sim_time: false
```

For the second node, run the following:

```
$ ros2 param dump /num_pub2
/num_pub2:
  ros__parameters:
    number: 4
    publish_period: 1.0
    start_type_description_service: true
    use_sim_time: false
```

As you can see, the output is exactly what you need to write inside a YAML file. You can then just copy and paste what you get in the terminal and create your own YAML file to load later (there's no need to set `use_sim_time` and `start_type_description_service`).

This `ros2 param dump` command can be useful for getting all parameters' values at once and for building a param YAML file quickly.

Setting a parameter's value from the terminal

Parameters are actually not set in stone for the entire life of a node. After you initialize a parameter's value with `ros2 run`, you can modify it from the terminal.

With our camera driver example, let's say you disconnect and reconnect the camera. The device name might change on Linux. If it were `/dev/ttyUSB0`, now it could be `/dev/ttyUSB1`. You could stop and start the node again with a different value for the device name parameter, but with the `ros2 param set` command, you could also just change the value directly while the node is still running.

To show you how it works, let's come back to our number application.

Stop all nodes and start one `number_publisher` node (here, I don't provide any parameter; we will use the default values):

```
$ ros2 run my_py_pkg number_publisher
```

Let's just verify the value for the `number` parameter:

```
$ ros2 param get /number_publisher number
Integer value is: 2
```

To modify a parameter from the terminal, you have to run `ros2 param set <node_name> <param_name> <new_value>`, as in the following example:

```
$ ros2 param set /number_publisher number 3
Set parameter successful
```

Of course, make sure to provide the correct data type for the parameter; otherwise, you will get an error. You can also load a YAML file directly with `ros2 param load <node_name> <yaml_file>` so that you can set several parameters at the same time:

```
$ ros2 param load /number_publisher ~/number_params.yaml
Set parameter number successful
Set parameter publish_period successful
```

After modifying a parameter, we check the parameter's value again:

```
$ ros2 param get /number_publisher number
Integer value is: 7
```

As you can see, the value was successfully changed. However, did this really work? Is the new parameter's value used in the code?

Let's verify that we are publishing the correct number:

```
$ ros2 topic echo /number
data: 2
---
```

Even if we changed the parameter's value, the new value was not updated inside the code. To do that, we will need to add a parameter callback. That's what we will see in a minute, but for now, let's just finish this section with the extra existing services that allow you to manage parameters.

Parameter services

If you remember, when we worked on services, you saw that for each node, we got an additional set of seven services, most of them related to parameters.

List all services for the `number_publisher` node:

```
$ ros2 service list
/number_publisher/describe_parameters
/number_publisher/get_parameter_types
/number_publisher/get_parameters
/number_publisher/get_type_description
/number_publisher/list_parameters
/number_publisher/set_parameters
/number_publisher/set_parameters_atomically
```

With those services, you can list parameters, get their value, and even set new values. Those services basically give you the same functionalities as the `ros2 param` command-line tool.

This is good news because getting and setting parameters from the terminal is not really practical and scalable in a real application. By using those services, you can create a service client in node A, which will get or modify parameters in node B.

I will not dive too far into this; you can experiment on your own with what you saw in *Chapter 6*. Let's just do a quick example here by modifying the `number` parameter. Let's first check which interface you need to use:

```
$ ros2 service type /number_publisher/set_parameters
rcl_interfaces/srv/SetParameters
```

Then, you can get more details with `ros2 interface show`. Finally, you can create a service client (inside a node) to modify a parameter. Let's do so from the terminal:

```
$ ros2 service call /number_publisher/set_parameters rcl_interfaces/
srv/SetParameters "{parameters: [{name: 'number', value: {type: 2,
integer_value: 3}}]}"
```

This is the same as running `ros2 param set /number_publisher number 3`. The benefit of the service is that you can use it inside any of your other nodes, with a service client from `rclpy` or `rclcpp`.

If you're wondering what `type: 2` means in the service request, here are all the types you can get or set with the parameter services:

```
$ ros2 interface show rcl_interfaces/msg/ParameterType
uint8 PARAMETER_NOT_SET=0
uint8 PARAMETER_BOOL=1
uint8 PARAMETER_INTEGER=2
uint8 PARAMETER_DOUBLE=3
uint8 PARAMETER_STRING=4
uint8 PARAMETER_BYTE_ARRAY=5
uint8 PARAMETER_BOOL_ARRAY=6
uint8 PARAMETER_INTEGER_ARRAY=7
uint8 PARAMETER_DOUBLE_ARRAY=8
uint8 PARAMETER_STRING_ARRAY=9
```

So, the number 2 corresponds to the `PARAMETER_INTEGER` type.

Now that you've seen how to set a parameter's value while the node is already running, let's continue with parameter callbacks. The problem so far is that if we modify a parameter, the value doesn't *reach* the code.

Updating parameters with parameter callbacks

After a parameter's value has been set when the node starts, you can modify it from the terminal or with a service client. To be able to receive the new value in your code, however, you will need to add what is called a parameter callback.

In this section, you will learn how to implement a parameter callback for Python and C++. This callback will be triggered whenever a parameter's value has been changed, and we will be able to get the new value in the code.

> **Note**
>
> You don't necessarily need to add parameter callbacks in your nodes. For some parameters, you will want to have a fixed value when you start the node and not modify this value anymore. Use parameter callbacks only if it makes sense to modify some parameters during the execution of a node.

Parameter callbacks are a great way to change a setting in your node without having to create yet another service. Let me explain that with the camera driver example. If you want to be able to change the device name while the node is running, the default way would be services. You would create a service server in your node that accepts requests to change the device name. However, doing this for each small setting in your node can be a hassle. With parameters, not only can you provide a different device name at runtime, but you can also modify it later by using the parameter services that each ROS 2 node already has. There's no need to make it more complicated than that.

Now, let's see how to solve the issue we had when setting a new value for the number parameter, and let's start with Python.

There are actually several parameter callbacks you could implement, but to keep things simple, I will just use one of them. Parameter callbacks are a nice and useful functionality, but it's not necessarily the most important when you begin with ROS 2. Thus, here you will get an overview of the functionality, and feel free to do more research on your own after finishing the book (you will find additional resources in *Chapter 14*).

Python parameter callback

Let's write our first Python parameter callback.

Open the number_publisher.py file and register a parameter callback in the node's constructor:

```
self.add_post_set_parameters_callback(self.parameters_callback)
```

We also add a new import line:

```
from rclpy.parameter import Parameter
```

Then, we implement the callback method:

```
def parameters_callback(self, params: list[Parameter]):
    for param in params:
        if param.name == "number":
            self.number_ = param.value
```

In this callback, you receive a list of Parameter objects. For each parameter, you can access its name, value, and type. With a for loop, we go through each parameter we get and set the corresponding values in the code. You could also decide to validate the values (for example, only accept positive numbers), but I will not do that here to keep the code minimal.

To make a quick test, run the `number_publisher` node again (no specified params; default values will be used). In another terminal, subscribe to the `/number` topic:

```
$ ros2 topic echo /number
data: 2
---
```

Now, change the parameter's value:

```
$ ros2 param set /number_publisher number 3
Set parameter successful
```

Let's now go back to the other terminal to observe the change:

```
$ ros2 topic echo /number
data: 3
---
```

The parameter's value has been changed, and we have received this value in the code, thanks to the parameter callback.

C++ parameter callback

The behavior for parameter callbacks in C++ is exactly the same as for Python. Let's have a look at the syntax. Open the `number_publisher.cpp` file and register the parameter callback in the constructor:

```
param_callback_handle_ = this->add_post_set_parameters_callback(
    std::bind(&NumberPublisherNode::parametersCallback, this, _1));
```

Here is the implementation for the callback:

```
void parametersCallback(
    const std::vector<rclcpp::Parameter> & parameters)
{
    for (const auto &param: parameters) {
        if (param.get_name() == "number") {
            number_ = param.as_int();
        }
    }
}
```

We get a list of `rclcpp::Parameter` objects. From this, we can check each parameter's name with the `get_name()` method. If the parameter's name matches, we get the value. Since we are receiving an integer here, we use the `as_int()` method. For a string, you would use the `as_string()` method, and so on. Please refer to the GitHub files for the complete code.

You have now seen the basics of parameter callbacks. You will not necessarily add them to all your nodes. They are great if you need to be able to modify a parameter's value after the node has been started.

Let's end this chapter with an additional challenge to make you practice more with parameters.

Parameter challenge

With this challenge, you will practice everything you've seen in this chapter: declaring and getting parameters in your code, providing parameters' values at runtime, and saving the values inside a YAML file. We will just skip parameter callbacks, but feel free to add them if you want to practice those too.

As usual for challenges, I will first explain what the challenge is and then provide the Python solution. You can find the complete code for both Python and C++ in the book's GitHub repository.

Challenge

We will continue to improve the `turtle_controller` node. For this challenge, we want to be able to choose different settings at runtime:

- Pen color on the right side
- Pen color on the left side
- Velocity to publish on the `cmd_vel` topic

To do that, you will add these parameters:

- `color_1`: Instead of just arbitrarily choosing a color for when the turtle is on the right side, we rename the color as `color_1`, and we get the value from a parameter. This parameter will be an integer list containing three values (`red`, `green`, `blue`).
- `color_2`: Same as for `color_1`, this one is the color used when the turtle is on the left side of the screen.
- `turtle_velocity`: By default, we used `1.0` and `2.0` for velocities sent on the `cmd_vel` topic. We make this a parameter so that we can provide the velocity at runtime. Instead of `1.0` and `2.0`, we will use `turtle_velocity` and `turtle_velocity * 2.0`.

To test this node, you will start the `turtle_controller` node with `ros2 run` and provide different values for the parameters. You should see if it works by watching how fast the turtle is moving and what the colors are for the pen. If needed, add some logs in the code to see what's happening.

As a last step for this challenge, you can put all the parameters inside a YAML file and load this YAML file at runtime.

Solution

Let's start by declaring the parameters that we will need for this challenge.

Open the `turtle_controller.py` file. Let's declare a few parameters at the beginning of the node's constructor:

```
self.declare_parameter("color_1", [255, 0, 0])
self.declare_parameter("color_2", [0, 255, 0])
self.declare_parameter("turtle_velocity", 1.0)
```

We provide default values that correspond to the same values we previously hardcoded. Thus, if we start the node without providing any parameters, the behavior will be the same as before.

After declaring the parameters, we can get their values:

```
self.color_1_ = self.get_parameter("color_1").value
self.color_2_ = self.get_parameter("color_2").value
self.turtle_velocity_ = self.get_parameter("turtle_velocity").value
```

We store the values inside class attributes so that we can reuse them later in the code.

> **Note**
>
> As a reminder, in Python, don't forget to add `.value` (without any parentheses) after `get_parameter()`. This is a common error that will lead to an exception when you start the node.

Then, we modify a few lines in the `callback_pose()` method:

```
if pose.x < 5.5:
    cmd.linear.x = self.turtle_velocity_
    cmd.angular.z = self.turtle_velocity_
else:
    cmd.linear.x = self.turtle_velocity_ * 2.0
    cmd.angular.z = self.turtle_velocity_ * 2.0
self.cmd_vel_pub_.publish(cmd)
```

Instead of hardcoding the velocity value, we use the one we got from the parameter.

Then, we set the pen color:

```
if pose.x > 5.5 and self.previous_x_ <= 5.5:
    self.previous_x_ = pose.x
    self.get_logger().info("Set color 1.")
    self.call_set_pen(
        self.color_1_[0],
        self.color_1_[1],
```

```
            self.color_1_[2]
        )
    elif pose.x <= 5.5 and self.previous_x_ > 5.5:
        self.previous_x_ = pose.x
        self.get_logger().info("Set color 2.")
        self.call_set_pen(
            self.color_2_[0],
            self.color_2_[1],
            self.color_2_[2]
        )
```

Here, we modify the logs so that they make more sense, as the color could be anything.

Finally, there are different ways you could pass the integer array to the `call_set_pen()` method. You could modify `call_set_pen()` so that it receives an array of three integers and extracts each number from it. Or, like I did here, you don't modify the method and you just make sure to pass the correct arguments.

The code is now finished. To test it, start the turtlesim node in one terminal and the turtle_controller node in another one. You can provide different values for the parameters. For example, if we want the velocity to be 1.5 and the colors to be black and white, we run the following:

```
$ ros2 run turtle_controller turtle_controller --ros-args -p
color_1:=[0,0,0] -p color_2:=[255,255,255] -p turtle_velocity:=1.5
```

You can also save those parameters inside a YAML file. Create a new YAML file (for example, in your home directory) named `turtle_params.yaml`. In this file, write this:

```
/turtle_controller:
  ros__parameters:
    color_1: [0, 0, 0]
    color_2: [255, 255, 255]
    turtle_velocity: 1.5
```

Then, you can start the turtle controller node with the YAML file directly:

```
$ ros2 run turtle_controller turtle_controller --ros-args --params-
file ~/turtle_params.yaml
```

That's it for this challenge. In the end, for each parameter, we did three things: we declared it, got its value, and used it in the code. This is not too complicated, and if you just know how to do that, you will be able to successfully handle parameters in your future ROS 2 applications.

Summary

In this chapter, you worked on parameters. Parameters allow you to provide settings for your nodes at runtime. Thus, with the same code, you could start several different nodes with different configurations. This increases the code reusability a lot.

To handle parameters in your nodes, follow these guidelines:

1. Declare the parameter so that it exists within the node. The best practice is to set a default value. This value will also set the type for the parameter.
2. Get the parameter's value and store it in your node—for example, in a private attribute.
3. Use this value in your code.

Then, when you start a node with `ros2 run`, you can specify any parameter's value you want.

You can also organize your parameters inside a YAML file, which makes it much more convenient when you start to have more than a handful of parameters. You will load the YAML file when you start a node.

Finally, you can also decide to allow parameters to be modified even after you've started a node. To do that, you will need to implement parameter callbacks.

Parameters make your nodes much more dynamic. In almost every node you run, you will have parameters. Using them makes it easier to scale your application by allowing different sets of configurations to be loaded.

Speaking of scaling, in the following chapter, we will dive into launch files. With launch files, you can start multiple nodes and parameters at once. This will be of great help when your application starts to grow.

9
Launch Files – Starting All Your Nodes at Once

At this point, you know how to write nodes, how to make them communicate with topics, services, and actions, and how to make them more dynamic with parameters.

In this last chapter of *Part 2*, we will bring everything together and go one step further toward making your application more scalable. Here, we will talk about launch files, which allow you to start all your nodes and parameters at once.

To start with launch files, it's important that you're comfortable with the concepts seen in the previous chapters. As a starting point, we will use the code inside the `ch8` folder from the book's GitHub repository (`https://github.com/PacktPublishing/ROS-2-from-Scratch`). You can find the final code for launch files in the `ch9` folder.

First, as always, I will use a real-life example to explain why you need launch files and what they are exactly. You will then dive into the code and create your own launch file with XML and Python (we will discuss which language is more appropriate). You will also experiment with extra configurations to fully customize your nodes inside launch files, and you will practice more with a final challenge.

By the end of this chapter, you will be able to properly scale your ROS 2 applications and know how to use or modify existing launch files. Almost every ROS 2 application or stack contains one or several launch files. Being comfortable with them is key to becoming a great ROS developer.

In this chapter, we will cover the following topics:

- What is a ROS 2 launch file?
- Creating and installing an XML launch file
- Creating a Python launch file – XML or Python for launch files?

- Configuring nodes inside a launch file
- Launch file challenge

What is a ROS 2 launch file?

After everything you've learned already, understanding the concept of launch files will not be very difficult.

You have experimented a bit with launch files in *Chapter 3*. We will now, as usual, start from scratch and see what a launch file is with an example. First, we'll look at why we need launch files.

Why launch files?

As your ROS 2 application starts to grow, so does the number of nodes and parameters. For example, a ROS stack I developed for a robotic arm had more than 15 nodes and 200 parameters. Imagine opening 15 terminals and starting all the nodes one by one with all the correct values for parameters. This would quickly become a nightmare.

For this explanation, let's assume we have the following nodes in our application:

- Three camera nodes with different settings
- Two LED panel nodes with varying numbers of LEDs
- One battery node
- Another node with more parameters

Here is what your application would look like:

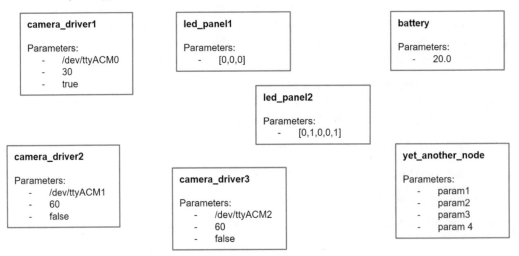

Figure 9.1 – ROS 2 application with seven nodes and sixteen parameters

To start all those nodes, you will need to open seven terminals and start the nodes one by one. For each node, you will also need to provide all the required parameters' values (with what you've seen in the previous chapter, you can use YAML param files to make this easier). Not only is this not scalable, but it will make your development process much slower and frustrating. With so many terminals, it's easy to make mistakes or to forget which terminal is doing what.

A solution you could think of is to create a script (a bash script, for example) to start all ros2 run commands from one file. That way, you could run your application from just one terminal. This would reduce development time and allow your application to scale.

Well, this is exactly what launch files are made for. There's no need to write your own script; all you need to do is create a launch file and follow a few syntax rules. Launch files can be installed within your ROS 2 application. Let's look at an example in the next section.

Example of a launch file with seven nodes

If we continue with our example, here is how your nodes would be organized:

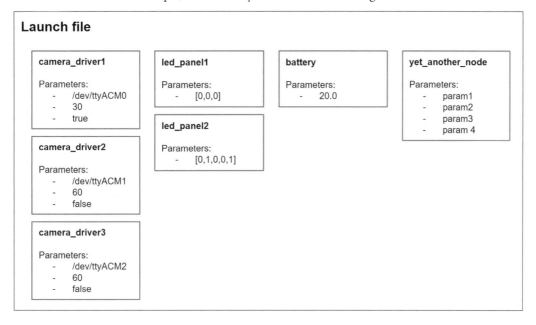

Figure 9.2 – Launch file with all nodes and parameters

Inside one file, you start all the nodes and provide the values you want for each parameter. This file can be written with XML, YAML, or Python—we will see how to do that in a moment. Then, once the launch file is written, you will install it (colcon build) and run it with the ros2 launch command-line tool.

It's not uncommon to have a few dozen nodes and a few hundred parameters inside one application. Without launch files, it would be impossible to quickly start the application, and you would spend most of your time debugging trivial things.

Launch files allow you to customize and scale your application easily. There is not much more to say; the concept is fairly straightforward. Most of the work is about learning how to implement one and knowing the features to customize your nodes to make them more dynamic. This is what we will dive into right now.

Creating and installing an XML launch file

You will now create your first launch file. We will start with XML. Later in this chapter, we will also write Python launch files and compare the two languages, but let's keep things simple to get started.

To properly create, install, and start a launch file, you need to do a bit of setup. In this section, we will follow all the necessary setup steps with a minimal launch file.

What we want to do here is to start the number application (`number_publisher` and `number_counter` nodes) from one terminal, with just one command line. Let's get started.

Setting up a package for launch files

Where should you put your launch files? You could theoretically create a launch file in any existing package.

However, this method can quickly lead to a dependency mess between packages. If package A requires package B, and you create a launch file in package B to start nodes from both packages, then you have created what's called a *dependency loop*. Package A depends on package B, and package B depends on package A. This is a very bad way to start a ROS application.

As a best practice, we will create a package dedicated to launch files. We will not modify any existing package; instead, we will create a completely independent one.

First, let's choose a name for this package. We will follow a common naming convention. We start with the name of the robot or application, followed by the `_bringup` suffix. As we don't have a robot here, we will call this package `my_robot_bringup`. If your robot were named *abc*, you would create an `abc_bringup` package.

Navigate to the `src` directory in your ROS 2 workspace and create this package. It will not contain any Python or C++ nodes. For the build type, you can choose `ament_cmake` (you could even omit the build type, as `ament_cmake` is the default anyway):

```
$ cd ~/ros2_ws/src/
$ ros2 pkg create my_robot_bringup --build-type ament_cmake
```

Alternatively, you could just run `$ ros2 pkg create my_robot_bringup`.

Once the package is created, we can remove directories that we don't need:

```
$ cd my_robot_bringup/
$ rm -r include/ src/
```

Then, we create a `launch` directory. This is where we will put all our launch files for this application:

```
$ mkdir launch
```

Before we create a launch file, let's finish the package configuration. Open the `CMakeLists.txt` file and add these lines:

```
find_package(ament_cmake REQUIRED)
install(DIRECTORY
  launch
  DESTINATION share/${PROJECT_NAME}/
)
ament_package()
```

This will install the `launch` directory when you build your package with `colcon build`.

Now, the package is correctly configured. You only need to do those steps once for each ROS 2 application. Then, to add a launch file, you just have to create a new file inside the `launch` folder. Let's do that.

Writing an XML launch file

Navigate to the `launch` folder you created inside the `my_robot_bringup` package. To create a launch file, you will first choose a name and then use the `.launch.xml` extension. Since we have named our application the *number app*, let's create a new file named `number_app.launch.xml`:

```
$ cd ~/ros2_ws/src/my_robot_bringup/launch/
$ touch number_app.launch.xml
```

Open the file, and let's start to write the content for the launch file.

First, you will need to open and close a `<launch>` tag. Everything you write will be between those two lines. This is the minimum code for an XML launch file:

```
<launch>
</launch>
```

Then, we want to start the `number_publisher` and `number_counter` nodes.

As a quick reminder, in the terminal, you would run this:

```
$ ros2 run my_py_pkg number_publisher
$ ros2 run my_cpp_pkg number_counter
```

Here, I started one node from the Python package and the other one from the C++ package. The two arguments we need to provide for `ros2 run` are the package name and executable name. This is the same inside a launch file. To add a node, use a `<node>` tag with the `pkg` and `exec` arguments:

```xml
<launch>
    <node pkg="my_py_pkg" exec="number_publisher"/>
    <node pkg="my_cpp_pkg" exec="number_counter"/>
</launch>
```

With this, we start the same two nodes from the launch file. As you can see, there's nothing very complicated. Later on in this chapter, we will see how to configure the application with remappings, parameters, namespaces, and so on. For now, let's focus on running this minimal launch file.

Installing and starting a launch file

You now have to install your new launch file before you can start using it.

As we are starting nodes from the `my_py_pkg` and `my_cpp_pkg` packages, we need to add the dependencies in the `package.xml` file of the `my_robot_bringup` package:

```xml
<exec_depend>my_py_pkg</exec_depend>
<exec_depend>my_cpp_pkg</exec_depend>
```

> **Note**
>
> Previously, we only used a `<depend>` tag when specifying dependencies. In this case, there is nothing to build; we only need the dependency when executing the launch file. Thus, we use a weaker tag, `<exec_depend>`.

For each new package you use in your launch files, you will add a new `<exec_depend>` tag in the `package.xml` file.

Now, we can install the launch file. To do so, you just need to build your package:

```
$ cd ~/ros2_ws/
$ colcon build --packages-select my_robot_bringup
```

Then, source your environment, and use the `ros2 launch` command-line tool to start the launch file. The full command is `ros2 launch <package_name> <launch_file_name>`:

```
$ ros2 launch my_robot_bringup number_app.launch.xml
```

You will see the following logs:

```
[INFO] [launch]: All log files can be found below /home/user/.ros/
log/...
[INFO] [launch]: Default logging verbosity is set to INFO
[INFO] [number_publisher-1]: process started with pid [21108]
[INFO] [number_counter-2]: process started with pid [21110]
[number_counter-2] [INFO] [1716293867.204728817] [number_counter]:
Number Counter has been started.
[number_publisher-1] [INFO] [1716293867.424510088] [number_publisher]:
Number publisher has been started.
[number_counter-2] [INFO] [1716293868.413350769] [number_counter]:
Counter: 2
[number_counter-2] [INFO] [1716293869.413321220] [number_counter]:
Counter: 4
[number_counter-2] [INFO] [1716293870.413321491] [number_counter]:
Counter: 6
```

What's happening here? Let's take a closer look:

1. A log file is created and the logging verbosity is set.
2. Each executable that you provided in the launch file is started as a new process. You can see the process name (for example, `number_publisher-1`) and the process ID (denoted as `pid`).
3. Then, as all nodes are started in the same terminal, you will see all logs from all nodes.

This example is quite simple, as we just start two executables with no additional configuration. Launch files will become quite handy when the number of nodes and settings gets bigger. Also, the `ros2 launch` command-line tool is very easy to use. There is not really much more than what we've seen here.

Now that you have completed the process to create, install, and start a launch file, let's talk about Python launch files.

Creating a Python launch file – XML or Python for launch files?

There are actually three languages you can use to create launch files in ROS 2: Python, XML, and YAML. I will not cover YAML launch files as they are not seldom used, and YAML doesn't have any competitive advantage over XML for launch files. Here, we will be focusing on Python and XML.

We will start this section by creating a Python launch file (the same application as before). Then, I will compare XML and Python launch files and give you some guidance on how to get the best out of them both.

Writing a Python launch file

As we already have a fully configured `my_robot_bringup` package for our application, there's no need to do anything else. All we have to do is create a new file inside the `launch` directory.

For Python launch files, you will use the `.launch.py` extension. Create a new file named `number_app.launch.py`. Here is the code required to start the `number_publisher` and `number_counter` nodes:

```python
from launch import LaunchDescription
from launch_ros.actions import Node

def generate_launch_description():
    ld = LaunchDescription()

    number_publisher = Node(
        package="my_py_pkg",
        executable="number_publisher"
    )

    number_counter = Node(
        package="my_cpp_pkg",
        executable="number_counter"
    )

    ld.add_action(number_publisher)
    ld.add_action(number_counter)
    return ld
```

The first thing you will notice is that the code is much, much longer than the XML one. I will come back to this in a minute when I compare Python and XML. For now, let's focus on the required steps to write a Python launch file:

1. The launch file must include a `generate_launch_description()` function. Make sure you don't make any typos.
2. In this function, you will need to create and return a `LaunchDescription` object. You can get this from the `launch` module.
3. To add a node in the launch file, you create a `Node` object (from `launch_ros.actions`) and specify the package and executable name. Then, you can add this object to the `LaunchDescription` object.

That's it for now, but there are more options that we will explore a bit later in this chapter.

Once you have written the launch file, make sure to add all required dependencies in the `package.xml` file of the `my_robot_bringup` package. As we already did that with the XML launch file (and we have the same dependencies here), we can skip this step.

Finally, to install this launch file, build the `my_robot_bringup` package again. Since we already wrote the necessary instructions in the `CMakeLists.txt` file, the launch file will be installed. All you need to do after that is to source your environment and start the launch file with `ros2 launch`:

```
$ ros2 launch my_robot_bringup number_app.launch.py
```

To create, install, and start a Python launch file, the process is the same as for an XML launch file. Only the code is different. Let's now compare the two languages regarding their use in launch files.

XML versus Python for launch files

I have a strong bias toward simplicity, so, from seeing the previous code examples, you can already guess where I'm going to stand.

To answer the XML versus Python question, let's first go back in time.

The issue with Python launch files

In ROS 1, the first version of ROS, XML was the only language used for launch files. Python was actually also available, but due to non-existent documentation, nobody knew about it.

At the beginning of ROS 2, the development team put a stronger emphasis on Python launch files and started to write the documentation only for Python, thus making it the default language for launch files. XML (and YAML) launch files were also supported, but again, due to non-existent documentation, nobody was using them.

I was initially enthusiastic about the idea of writing Python launch files, as this meant you could take advantage of the Python logic and syntax to make launch files much more dynamic and easier to write. That's the theory, but in practice, I realized I didn't see any programming logic in most of the launch files I found, and it was just another—more complex and difficult—way to write a description, which is basically why XML exists in the first place.

You can already see the added complexity in the two previous examples. To start two nodes, it takes four lines in XML and twenty lines in Python (I could optimize the code and make it less than fifteen lines, but that's still a lot more). For the same number of nodes, you can expect Python launch files to be two to five times longer than the XML version.

Also, with more functionalities (parameters, arguments from the terminal, conditions, paths, and so on), you will have to use an increasing amount of Python imports that are hard to find and use. You will realize this as we see more examples of XML and Python launch files all along this book.

Fortunately, XML is coming back, as the official documentation is starting to include it as well as Python. More and more developers have started to use XML launch files again, which is a good thing because more online tutorials and open source code will include them.

How to combine XML and Python launch files in your application

XML launch files are much simpler and smaller to write than Python launch files. However, for some advanced use cases, Python will be the only choice, as it contains some functionalities that are not available for XML. This could be a problem because if you need just one Python functionality, it would mean that you'd need to write the entire launch file in Python.

Fortunately, there is a very easy way to solve that. As we will see in a minute, you can include any kind of launch file into any other launch file, be it Python, XML, or YAML.

So, if you absolutely need to use Python for a specific launch functionality, then go ahead and create a Python launch file for that. You can then include this launch file in your *main* XML launch file. You can also include any other existing Python launch file (from an already installed package) that contains the functionality you need. By doing this, you keep your code minimal and simple.

Now, what to do when you need to create a Python launch file for a specific use case? The syntax is really complicated, and there are too many imports for any functionality. It can quickly become a challenge.

What I myself do when I have to create a Python launch file is to try to find an existing launch file on GitHub that does what I want and tweak the code so that it works with my application. I gave up on trying to learn or even memorize the Python launch file syntax. I am not usually a fan of the "copy/paste from the internet" method, but I make an exception for Python launch files.

In the end, it's a matter of choice for you. A correctly written XML, YAML, or Python launch file will do the exact same thing. As for YAML, it's just another markup language, and I find XML easier to use for launch files. My recommendation is to use XML whenever possible. Use Python only if you have to and only for the functionalities that require Python. Then, include the Python launch file inside your XML one.

Following this process will make your life easier when developing ROS 2 applications.

Including a launch file inside another launch file

Since I talked about including a Python launch file inside an XML launch file, let's see how to do that. The syntax won't be that complicated.

Make sure you add everything inside `<launch></launch>` tags. To include another launch file, use an `<include>` tag. Here is an example:

```
<launch>
    <include file="$(find-pkg-share
        my_robot_bringup)/launch/number_app.launch.py" />
</launch>
```

This line, with `find-pkg-share`, will find the path to the `number_app.launch.py` launch file inside the `my_robot_bringup` package. Then, the content of the launch file will be included. Even if you include a Python launch file inside an XML one, this will work.

You can reuse this line in any other XML launch file; just replace the package name and launch filename.

Now, if you wanted to do the opposite (which means including an XML launch file inside a Python launch file), here is what you would need to write:

```
from launch import LaunchDescription
from launch.actions import IncludeLaunchDescription
from launch_xml.launch_description_sources import
XMLLaunchDescriptionSource
import os
from ament_index_python import get_package_share_directory

def generate_launch_description():
    ld = LaunchDescription()
    other_launch_file = IncludeLaunchDescription(
     XMLLaunchDescriptionSource(os.path.join(
       get_package_share_directory('my_robot_bringup'),
                    'launch/number_app.launch.xml')))
    ld.add_action(other_launch_file)
    return ld
```

This code example illustrates what I was saying about the extra complexity brought by Python launch files. This complexity is not justified here, as it adds nothing compared to the XML file.

With those two code examples, you can now combine any XML and Python launch files.

Now that you have seen the process of creating a launch file in both XML and Python, let's go a bit further and add some extra configuration for the nodes.

Configuring nodes inside a launch file

So far, we have just started two nodes, with zero extra configuration. When you start a node with `ros2 run`, as we have seen in the previous chapters in *Part 2*, you can rename it, rename topics/services/actions, add parameters, and so on.

In this section, you will learn how to do that inside a launch file. We will also introduce the concept of namespaces. All code examples will be in XML and Python.

Renaming nodes and communications

In an XML launch file, to rename a node, simply add a `name` argument in a `<node>` tag:

```
<node pkg="your_package" exec="your_exec" name="new_name" />
```

Changing the name for a topic/service/action is actually named *remapping*. To remap a communication, you have to use a `<remap>` tag, inside the `<node>` tag:

```
<node pkg="your_package" exec="your_exec">
    <remap from="/topic1" to="/topic2" />
</node>
```

You can add as many `<remap>` tags as you want, each one in a new line.

> **Note**
> This is a quick XML reminder, but it can be useful if you're not used to XML and can prevent lots of errors in the future. For one-line tags, you open the tag and end it with `/>` (for example, `<node />`). If you need to add a tag inside a tag, you then have to open the tag and close it later, like we did for `<launch>...</launch>` or `<node>...</node>`.

From this, let's say we want to start two `number_publisher` nodes and one `number_counter` node. On top of that, we also want to remap the topic from `number` to `my_number`. Here is the full XML launch file:

```
<launch>
    <node pkg="my_py_pkg" exec="number_publisher" name="num_pub1">
        <remap from="/number" to="/my_number" />
    </node>
    <node pkg="my_py_pkg" exec="number_publisher" name="num_pub2">
        <remap from="/number" to="/my_number" />
    </node>
    <node pkg="my_cpp_pkg" exec="number_counter">
        <remap from="/number" to="/my_number" />
    </node>
</launch>
```

We rename the two `number_publisher` nodes to avoid name conflicts. Then, we make sure to add the same `<remap>` tag for all nodes in which we use a publisher or subscriber on the `number` topic.

> **Additional tip**
> When you rename nodes and remap communications, use `rqt_graph` to verify that everything is working fine. With the graphical view, you can easily spot if a topic name is not the same on both sides of the communication.

Here is the code to do the same thing with a Python launch file:

```python
from launch import LaunchDescription
from launch_ros.actions import Node

def generate_launch_description():
    ld = LaunchDescription()

    number_publisher1 = Node(
        package="my_py_pkg",
        executable="number_publisher",
        name="num_pub1",
        remappings=[("/number", "/my_number")]
    )
    number_publisher2 = Node(
        package="my_py_pkg",
        executable="number_publisher",
        name="num_pub2",
        remappings=[("/number", "/my_number")]
    )
    number_counter = Node(
        package="my_cpp_pkg",
        executable="number_counter",
        remappings=[("/number", "/my_number")]
    )

    ld.add_action(number_publisher1)
    ld.add_action(number_publisher2)
    ld.add_action(number_counter)
    return ld
```

After renaming and remapping, let's see how to add parameters to your nodes inside a launch file.

Parameters in a launch file

Setting parameters' values for a node in a launch file is pretty straightforward. We will first see how to provide the values directly, and then how to load a YAML file.

Setting parameters' values directly

To add a parameter's value for a node in an XML launch file, you first need to open and close the `<node></node>` tag. Inside this tag, you will add one `<param>` tag per parameter, with two arguments: `name` and `value`.

Here is an example, where we set the `number` and `publish_period` parameters for the `number_publisher` node:

```
<node pkg="my_py_pkg" exec="number_publisher">
    <param name="number" value="3" />
    <param name="publish_period" value="1.5" />
</node>
```

It will work the same as adding `-p <parameter>:=<value>` after the `ros2 run` command.

Now, you can combine renaming, remapping, and setting parameters. Let's add parameters to the previous example:

```
<node pkg="my_py_pkg" exec="number_publisher" name="num_pub1">
    <remap from="/number" to="/my_number" />
    <param name="number" value="3" />
    <param name="publish_period" value="1.5" />
</node>
```

In a Python launch file, you need to add a list of dictionaries in the `Node` object:

```
number_publisher1 = Node(
    package="my_py_pkg",
    executable="number_publisher",
    name="num_pub1",
    remappings=[("/number", "/my_number")],
    parameters=[
        {"number": 3},
        {"publish_period": 1.5}
    ]
)
```

Setting each parameter's value like this will work fine if you only have a handful of parameters. For bigger numbers, it's more suitable to use a YAML file.

> **Note**
> Do not confuse YAML param files with YAML launch files. Launch files can be written in Python, XML, and YAML (though we didn't use YAML in this book). Any of those launch files can include YAML param files, to add parameters' values for the nodes in the launch file.

Installing and loading a YAML param file in a launch file

To provide parameters' values using a YAML file, you will need to follow this process:

1. Create a YAML file with the values.
2. Install this file inside the `_bringup` package.
3. Load the YAML file in your launch file (we will do that with XML and then Python).

For this example, we are going to reuse the `number_params.yaml` file that we created in *Chapter 8*. In this file, you can find the following code:

```
/num_pub1:
  ros__parameters:
    number: 3
    publish_period: 0.5

/num_pub2:
  ros__parameters:
    number: 4
    publish_period: 1.0
```

This will perfectly match the nodes that we launched in the previous example, as the names are exactly the same.

Now, what we have done so far is just provide the path to the file when starting a node with `ros2 run`. To use the YAML param file inside a launch file, we will need to install it in the package.

To do that, create a new directory inside the `my_robot_bringup` package. You could choose any name for that directory, but we will follow a common convention and name it `config`:

```
$ cd ~/ros2_ws/src/my_robot_bringup/
$ mkdir config
```

Put the `number_params.yaml` file inside this `config` directory. This is where you will also put all other YAML param files for this application.

Now, to write instructions to install this directory (and all the YAML files inside), open the `CMakeLists.txt` file of the `my_robot_bringup` package and add one line:

```
install(DIRECTORY
  launch
  config
  DESTINATION share/${PROJECT_NAME}/
)
```

You only need to do this once. Any other file inside the `config` directory will be installed when running `colcon build` for that package.

Before we build the package, let's modify the launch file so that we can use this YAML param file. The way to do this in XML is easy. You will add a `<param>` tag, but instead of `name` and `value`, you need to specify a `from` argument:

```
<node pkg="my_py_pkg" exec="number_publisher" name="num_pub2">
    <remap from="/number" to="/my_number" />
    <param from="$(find-pkg-share
            my_robot_bringup)/config/number_params.yaml" />
</node>
```

As we've seen previously in this chapter, `$(find-pkg-share <package_name>)` will locate the installation folder for that package. Then, you only need to finish with the relative path to the file you want to retrieve.

To test this, first build your package. This will install the YAML param files and the launch files. Then, source your environment and start the XML launch file.

That's it for parameters. Let's now see the Python version. In your launch file, add the following imports:

```
from ament_index_python.packages import get_package_share_directory
import os
```

Then, retrieve the YAML file:

```
param_config = os.path.join(
    get_package_share_directory("my_robot_bringup"),
    "config", "number_params.yaml")
```

Finally, load the configuration into the node:

```
number_publisher2 = Node(
    ...
    parameters=[param_config]
)
```

With this, you should be able to start any node you want with any number of parameters without having any scaling issues.

Let's now finish this section with namespaces. I have briefly mentioned them a few times during this book. As you now have a better understanding of how names work in ROS 2, and as namespaces are especially useful in launch files, this is a good time to start with them.

Namespaces

Namespaces are quite common in programming, and you are probably already familiar with them. With a namespace, you can group some functionalities (variables, functions, and so on) inside one *container* that has a name. This can help you better organize your code and avoid name conflicts.

In ROS, namespaces are also quite practical. Let's say you want to start an application that contains two identical robots, but you want to be able to control each robot independently. Instead of renaming the nodes, topics, services, and actions for each robot, you could just add a namespace.

If you have a node named `robot_controller` and a topic named `cmd_vel`, then those can become `/robot1/robot_controller` and `/robot1/cmd_vel` for the first robot. For the second robot, this would be `/robot2/robot_controller` and `/robot2/cmd_vel`. This way, the two robots are still running on the same application, but you make sure that the velocity command for each robot is independent.

As you make progress with ROS 2 and learn new stacks and plugins, you will encounter namespaces everywhere. Let's now see how to work with namespaces. As we have not done this previously, we will first use namespaces with the `ros2 run` command line, and then add them in our launch file.

Starting a node inside a namespace

Adding a namespace to a node is quite straightforward.

First of all, after the `ros2 run <package> <executable>` command, you add `--ros-args` once. Then, to specify a namespace, you will write `-r __ns:=<namespace>`. The `-r` option (or `--remap`) is the same as the one for renaming a node, only instead of `__node`, you use `__ns` here.

Let's start our `number_publisher` node inside a `/abc` namespace:

```
$ ros2 run my_py_pkg number_publisher --ros-args -r __ns:=/abc
[INFO] [1716981935.646395625] [abc.number_publisher]: Number publisher
has been started.
```

After this, you can check what the node and topic names are:

```
$ ros2 node list
/abc/number_publisher
$ ros2 topic list
/abc/number
/parameter_events
/rosout
```

As you can see, `/abc` was added to the node name but also to the topic name—if you have services and actions, the namespace will be equally applied.

> **Important note**
>
> The namespace was successfully applied because the topic name defined in the code is `number` without any leading slash. If you had written `/number` in the code, then the topic would have been considered to be in the *global* scope or namespace. Adding a namespace to the node will change the node name but not the topic name. Thus, pay attention to this when defining communication (topic, service, action) names in your code.

Now, as the topic name is `/abc/number`, if we want to start the `number_counter` node and receive some data, we need to either rename the topic or also add a namespace to the node:

```
$ ros2 run my_cpp_pkg number_counter --ros-args -r __ns:=/abc
[abc.number_counter]: Number Counter has been started.
[abc.number_counter]: Counter: 2
[abc.number_counter]: Counter: 4
```

When adding namespaces, name mismatches can become a frequent issue. One of the best ways to verify that things are working is to run `rqt_graph`:

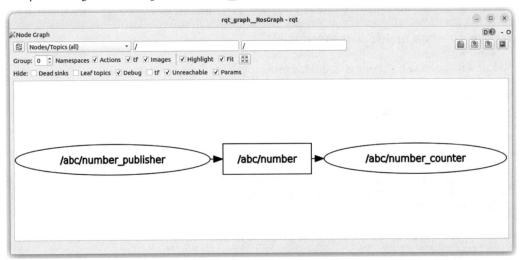

Figure 9.3 – Double-checking namespaces with rqt_graph

With this, you can see that both nodes are publishing or subscribing to the `/abc/number` topic.

> **Note**
>
> You can combine any type of renaming. For example, you could both add a namespace and rename the node: `$ ros2 run my_py_pkg number_publisher --ros-args -r __ns:=/abc -r __node:=num_pub`.

Now that you know how to provide a namespace for a node at runtime, let's see how to do this inside a launch file.

Specifying a namespace in a launch file

To add a namespace to a node in an XML launch file, you just have to add a `namespace` argument inside the `<node>` tag. Let's continue with our previous example:

```
<node pkg="my_py_pkg" exec="number_publisher" name="num_pub1"
namespace="/abc">
```

For Python, the syntax is also quite easy; here too, you just need to add a `namespace` argument inside the `Node` object:

```
number_publisher1 = Node(
    package="my_py_pkg",
    executable="number_publisher",
    namespace="/abc",
    name="num_pub1",
    ...
```

If you add a namespace to this node, you will also add the same namespace to nodes that are directly communicating with it:

```
<node pkg="my_py_pkg" exec="number_publisher" name="num_pub1"
namespace="/abc">
...
<node pkg="my_cpp_pkg" exec="number_counter" namespace="/abc">
```

Adding namespaces to nodes in a launch file is quite straightforward. However, there is one important thing you need to pay attention to. If you are using YAML param files, you also need to specify the namespace in the YAML file. Open the `number_params.yaml` file and add the namespace to the node name:

```
/abc/num_pub2:
  ros__parameters:
    number: 4
    publish_period: 1.0
```

If you don't do this, the parameters will be applied to the `/num_pub2` node, which doesn't exist, since it's named `/abc/num_pub2`. This can be a common source of errors, so make sure you double-check param files when adding namespaces.

After all those modifications, make sure to build the `my_robot_bringup` package again and source the environment before you start any launch file.

You have now seen a few ways to configure your nodes inside a launch file. With this base knowledge, you can already scale your application a lot. Let's finish this chapter with a new challenge so that you can practice more on your own.

Launch file challenge

In this challenge, you will practice more with launch files, YAML param files, remappings, and namespaces. This will be the conclusion of *Part 2*. To complete this challenge, you can decide to write the launch file in XML, Python, or both.

Challenge

What we want to do here is to start two `turtlesim` windows, each one with one turtle. Then, for each turtle, we run a `turtle_controller` node (the one we have been developing throughout the previous chapters).

The goal is to have each `turtle_controller` node controlling only one turtle. This is what the result should look like:

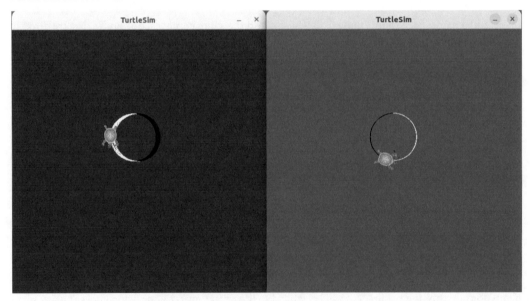

Figure 9.4 – Two different turtles with two independent controllers

For each turtle, we will apply different settings (parameters):

- First `turtlesim` window:

    - **Background color**: Red (you can pick `128` for the RGB value)

- First controller:
  - **Color 1**: Black
  - **Color 2**: White
  - **Velocity**: 1.5
- Second `turtlesim` window:
  - **Background color**: Green (you can pick 128 for the RGB value)
- Second controller:
  - **Color 1**: White
  - **Color 2**: Black
  - **Velocity**: 0.5

Here are the steps you can take:

1. Create a `turtle_params.yaml` file with the parameters for each node. Install this in the `my_robot_bringup` package.
2. Create a new launch file and start the four nodes. Load the parameters from the YAML param file. Put the different nodes into appropriate namespaces (to keep it simple, use `t1` and `t2` for `turtle1` and `turtle2`, respectively).
3. Build, source, and start the launch file. You will see that some topics and services are not matching, and thus you will know what remappings you need to add.

To make it easier, start with just one pair of nodes (`turtlesim` and `turtle_controller`), and then add another pair when it's working.

Here is an important point for this challenge: we will not modify any of the existing code—even if it would make things easier. The goal is to take the nodes exactly as they are (use the code from the `ch8` folder in the repo) and make things work using appropriate namespaces and remappings in the launch file and YAML param file.

Solution

Create a new file named `turtle_params.yaml`, inside the `config` directory of the `my_robot_bringup` package. As a base, you can take the one that we did in the parameter challenge for *Chapter 8*.

In this file, we will add parameters for all four nodes. Before we do this, we need to know exactly what will be the name for each node, including the namespaces.

With the `t1` and `t2` namespaces, if we just add a namespace and we don't rename the nodes, we will then have these names:

- `/t1/turtlesim`
- `/t2/turtlesim`
- `/t1/turtle_controller`
- `/t2/turtle_controller`

After making this choice, we can write the YAML param file:

```
/t1/turtlesim:
  ros__parameters:
    background_r: 128
    background_g: 0
    background_b: 0

/t2/turtlesim:
  ros__parameters:
    background_r: 0
    background_g: 128
    background_b: 0

/t1/turtle_controller:
  ros__parameters:
    color_1: [0, 0, 0]
    color_2: [255, 255, 255]
    turtle_velocity: 1.5

/t2/turtle_controller:
  ros__parameters:
    color_1: [255, 255, 255]
    color_2: [0, 0, 0]
    turtle_velocity: 0.5
```

This contains all the configurations given in the challenge. Now, create a new launch file (for example, `turtlesim_control.launch.xml`) inside the `launch` directory.

In this launch file, let's start with something simple. We want to try to run one `turtlesim` node and one `turtle_controller` node, using the `t1` namespace:

```xml
<launch>
    <node pkg="turtlesim" exec="turtlesim_node" namespace="t1">
        <param from="$(find-pkg-share
            my_robot_bringup)/config/turtle_params.yaml" />
    </node>
```

```xml
    <node pkg="turtle_controller" exec="turtle_controller"
namespace="t1">
        <param from="$(find-pkg-share
            my_robot_bringup)/config/turtle_params.yaml" />
    </node>
</launch>
```

As we are starting nodes from the `turtlesim` and `turtle_controller` packages, we also add two new `<exec_depend>` tags in the `package.xml` file:

```xml
<exec_depend>turtlesim</exec_depend>
<exec_depend>turtle_controller</exec_depend>
```

Now, if you launch this (make sure to build and source first), you will see the `turtlesim` node, but the turtle won't move. Why is that?

If you look at the topic list, you will find these two topics:

```
$ ros2 topic list
/t1/turtle1/cmd_vel
/turtle1/cmd_vel
```

With `rqt_graph`, you can also see that the `turtlesim` node is subscribing to `/t1/turtle1/cmd_vel`, but the `turtle_controller` node is publishing on `/turtle1/cmd_vel`. Why did the namespace work for the node name but not for the topic name?

This is because we wrote `/turtle1/cmd_vel` in the code, and not `turtle1/cmd_vel`. The fact that we added a leading slash makes the namespace the *global* namespace. Thus, if you try to add a namespace to that, it will not be taken into account.

We have two options here: either we modify the code (we simply need to remove this leading slash) or we adapt the launch file to make this work. As specified in the challenge instructions, we are not going to modify the code. The reason why I'm adding this constraint is because, in real life, you won't necessarily be able to modify the code of the nodes you run. Thus, knowing how to solve a name mismatch without touching the code is a great skill to have.

So, if you look at the topic and service names (we don't use actions here), you will see that we have two topics and one service to modify. Let's add some `<remap>` tags inside the node:

```xml
<node pkg="turtle_controller" exec="turtle_controller" namespace="t1">
    <param from="$(find-pkg-share
        my_robot_bringup)/config/turtle_params.yaml" />
    <remap from="/turtle1/pose" to="/t1/turtle1/pose" />
    <remap from="/turtle1/cmd_vel" to="/t1/turtle1/cmd_vel" />
    <remap from="/turtle1/set_pen" to="/t1/turtle1/set_pen" />
</node>
```

You can now start the launch file, and you will see the turtle moving. Now that we have this working, adding a second pair of nodes is easy. We basically need to copy/paste the two nodes and replace `t1` with `t2`:

```xml
<node pkg="turtlesim" exec="turtlesim_node" namespace="t2">
    <param from="$(find-pkg-share
            my_robot_bringup)/config/turtle_params.yaml" />
</node>

<node pkg="turtle_controller" exec="turtle_controller" namespace="t2">
    <param from="$(find-pkg-share
            my_robot_bringup)/config/turtle_params.yaml" />
    <remap from="/turtle1/pose" to="/t2/turtle1/pose" />
    <remap from="/turtle1/cmd_vel" to="/t2/turtle1/cmd_vel" />
    <remap from="/turtle1/set_pen" to="/t2/turtle1/set_pen" />
</node>
```

The challenge is now complete. If you start this launch file, you will see two `turtlesim` windows, each one containing a turtle that moves at a different speed and using different pen colors.

You can find the complete code and package organization in the book's GitHub repository (including the Python launch file).

Summary

In this chapter, you worked on ROS 2 launch files. Launch files allow you to properly scale your application with multiple nodes, parameters, and sets of configuration.

You can write a launch file in Python, XML, or YAML. Here, you discovered the Python and XML syntax and saw that XML is probably the best choice by default. The syntax is much easier, and the code is much shorter. If you ever need to combine XML and Python launch files, you can do so by including a launch file in another one.

The best practice is to set up a dedicated package for launch files and YAML files. You can name the package using the `_bringup` suffix. Launch files will be installed in a `launch` folder, and YAML param files in a `config` folder.

If you correctly understand how to start nodes with the `ros2 run` command, then doing so in a launch file is pretty straightforward: you just need to provide the package and executable name for each node. The only thing to learn is the XML or Python syntax.

In a launch file, you can also configure your nodes in multiple ways:

- Renaming the node and/or adding a namespace
- Remapping topics, services, and actions
- Adding parameters, individually or from a YAML param file

This is what we have seen so far, but there are many other ways to configure your nodes that you will discover throughout your ROS 2 learning journey.

Part 2 of this book is now finished. You have discovered all the core concepts that will allow you to write complete ROS 2 applications and join existing ROS 2 projects. You should now be able to interact with any ROS 2 node, write code to communicate with it, and scale your application with parameters and launch files.

Now, this part was heavily focused on programming (Python and C++), which is incredibly important, but ROS 2 is more than just that. In *Part 3*, we will dive into some additional concepts and tools (**TransForms (TFs)**, **Unified Robot Description Format (URDF)**, **Gazebo**) so that you can design a custom application for a robot, including a 3D simulation. This, combined with the programming we did in *Part 2*, will be the backbone of any ROS 2 application you work on.

Part 3: Creating and Simulating a Custom Robot with ROS 2

In this third and last part, you will go beyond just the Python or C++ code. New concepts and tools will be introduced, such as TF, URDF, RViz, and Gazebo. With those, and the knowledge you got in *Part 1* and *Part 2*, you will build a new project to simulate a robot with ROS 2. Finally, the last chapter concludes the book with recommendations on what to do and learn next.

This part contains the following chapters:

- *Chapter 10, Discovering TFs with RViz*
- *Chapter 11, Creating a URDF for a Robot*
- *Chapter 12, Publishing TFs and Packaging the URDF*
- *Chapter 13, Simulating a Robot in Gazebo*
- *Chapter 14, Going Further – What to Do Next*

10
Discovering TFs with RViz

In *Part 3* of this book, you will create a robot simulation with ROS 2. However, before you get started, you first need to understand what **TransForms** (**TFs**) are.

In ROS, a TF is the transformation between two frames in 3D space. TFs will be used to track the different coordinate frames of a ROS robot (or system with multiple robots) over time. They are used everywhere and will be the backbone of any robot you create.

To understand TFs, we will first look at an existing robot model. As we did back in *Chapter 3*, here, we will discover the concepts by experimenting, and you will build an intuition of how things work. During this phase, you will discover a few new ROS tools, including **RViz**, a 3D visualization package.

You will see for yourself how TFs work, how they are related to each other, and how to visualize the TF tree for any robot. By the end of this chapter, you will understand what TFs are, what problems they solve, and how they are used in a ROS 2 application.

Good news: once you understand TFs, well, it's the same principle for any ROS robot, so you can directly apply what you learn here to your future projects.

This chapter will be quite small and quick to finish. We won't write any code here, and there is no GitHub repository. All you have to do is follow the experiments. Not everything has to make sense right now; the goal is to get enough context to understand what we will be doing later. Don't hesitate to come back to this chapter once you've finished *Part 3*.

In this chapter, we will cover the following subjects:

- Visualizing a robot model in RViz
- What are TFs?
- Relationship between TFs
- What problem are we trying to solve with TFs?

Technical requirements

At the beginning of the book, I gave you two options: either installing Ubuntu with a dual boot or in a virtual machine.

If you chose the VM path, you should have been fine for all chapters in *Part 1* and *Part 2*. For this chapter and the next two, we will use a 3D visualization tool (RViz) that might not work if your computer is not powerful enough.

I would suggest first trying to run the commands from this chapter. If it doesn't work well (it's too slow, for example), then I strongly recommend you set up a dual boot with Ubuntu and ROS 2 (see the instructions in *Chapter 2*). If RViz works fine, then continue like this for now. The dual boot will be required for *Chapter 13*.

Visualizing a robot model in RViz

In this section, you will discover RViz. RViz allows you to visualize a robot model in 3D and contains many plugins and functionalities that will help you develop your robotics applications. With RViz, you will be able to visualize the TFs for a robot, so we can start to understand what they are.

As we haven't created a robot model yet, we will use one from an existing ROS 2 package named `urdf_tutorial`. We will load a robot model in RViz and learn how to navigate the software.

Let's start by setting up everything we need for this chapter.

Installation and setup

First of all, there is no need to install RViz. It was already included when you installed ROS 2 at the beginning of the book (with the `sudo apt install ros-<distro>-desktop` command).

To visualize TFs for a robot model on RViz, we will install a new ROS package named `urdf_tutorial`. This package contains some existing launch files and robot model files (how to create a robot model will be the focus of the next chapter).

If you remember, to install a ROS 2 package with `apt`, you have to start with `ros`, then write the name of the distribution you are using, and finally, add the package name. All words are separated with dashes (not underscores).

Open a terminal and install this package:

```
$ sudo apt install ros-<distro>-urdf-tutorial
```

Then, so that you can use the package, make sure you source the environment or simply open a new terminal.

Let's now visualize a robot model.

Starting RViz with a robot model

The `urdf_tutorial` package contains a launch file, named `display.launch.py`, that will start RViz and load a robot model into it. For now, we will just use it, and in the following chapters, we will understand how this process works so we can replicate it.

So, we need to start this launch file and also load a robot model. Where will we get one? There are some existing models in the `urdf_tutorial` package. To find them, navigate to the `share` directory where the package was installed, and you will find a `urdf` folder under the package name:

```
$ cd /opt/ros/<distro>/share/urdf_tutorial/urdf/
```

A **Unified Robot Description Format** (**URDF**) file is basically the description of a robot model. We will come back to this in the next chapter. For now, all we want to do is to visualize one. In the `urdf` folder, you can find several robot model files:

```
$ ls
01-myfirst.urdf          04-materials.urdf        07-physics.urdf
02-multipleshapes.urdf   05-visual.urdf           08-macroed.urdf.xacro
03-origins.urdf          06-flexible.urdf
```

You can now start a robot model in RViz by launching the `display.launch.py` file and add the path to the robot model with an additional `model` argument after the launch file:

```
$ ros2 launch urdf_tutorial display.launch.py model:=/opt/ros/<distro>/share/urdf_tutorial/urdf/07-physics.urdf
```

> **Note**
> To avoid errors, it's better to provide the absolute path to the .urdf file, even if you run the command from the same folder.

After running the command, you should see something like this:

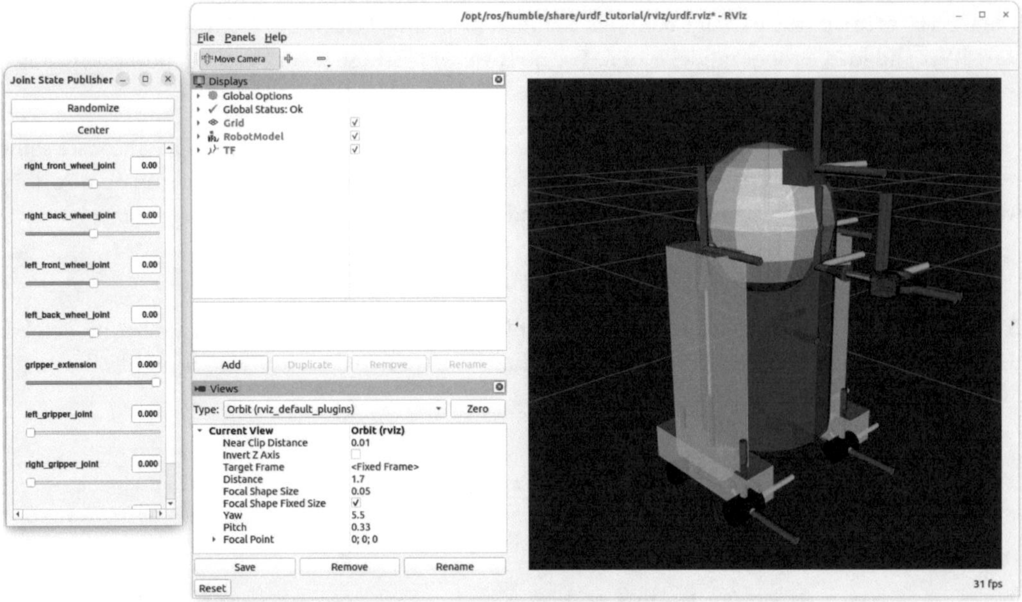

Figure 10.1 – A robot model on RViz

You will get two windows: the main one (**RViz**) with the robot model, and a **Joint State Publisher** window with some cursors. The robot model we have here is a replica of a famous science-fiction movie robot. It has some wheels, a torso, a head, and a gripper.

Let's focus on the main window (**RViz**) for now. Take some time to learn how to navigate in the 3D space and move around the robot. You can use the left click, right click, and mouse wheel. For this, it's best to have a mouse, but you could still manage to navigate with the touchpad of a laptop, although it's less ergonomic.

You can also resize the window and the various sections inside RViz. Pretty much everything you see can be customized. Now that you can load a robot model in RViz, we will start to experiment with TFs.

What are TFs?

There are two main parts in a robot model: links and TFs. In this section, we will visualize them both and understand how they work together.

Let's start with links.

Links

Have a look at the menu on the left side of the **RViz** window. There, you will see, in blue bold letters, **RobotModel** and **TF**. This is what we will focus on in this chapter. As you can see, you can enable or disable both menus.

Disable **TF**, keep **RobotModel**, and expand the menu. There, you can find a submenu called **Links**.

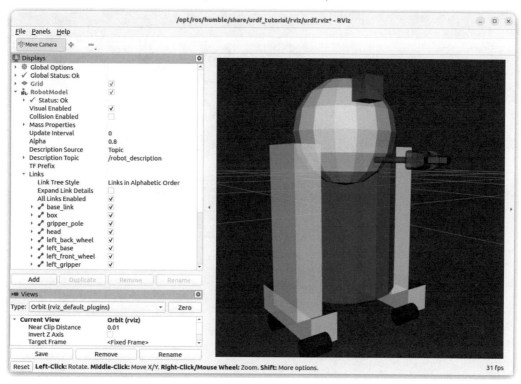

Figure 10.2 – RobotModel with Links menu on RViz

Check and uncheck some boxes. As you can see from this menu, a *link* is one rigid part (meaning one solid part with no articulation) of the robot. Basically, in ROS, a robot model will consist of a collection of rigid parts put together.

In this example, links are represented by basic shapes: boxes, cylinders, and spheres. Those rigid parts do nothing on their own, so how are they connected, and how do they move between each other?

This is where we introduce TFs.

TFs

Let's now check the **TF** box. You can keep **RobotModel** checked or unchecked. Inside the **TF** menu, you have a submenu called **Frames**, and you can also enable or disable each frame for the robot.

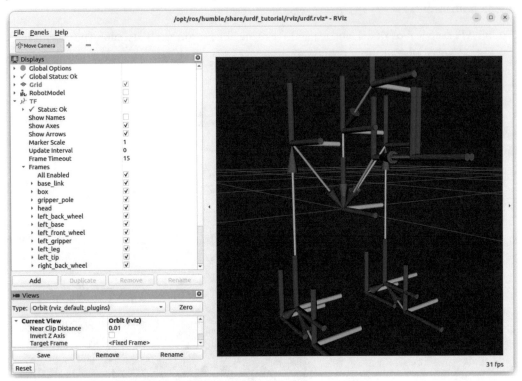

Figure 10.3 – Frames and TFs on RViz

The axes you see here (red, green, and blue coordinate systems) represent the frames, or basically the origin of each link of the robot.

Coordinate systems follow the right-hand rule in ROS. Following *Figure 10.4*, you have the following:

- X axis (red) pointing forward
- Y axis (green) pointing 90 degrees to the left
- Z axis (blue) pointing up

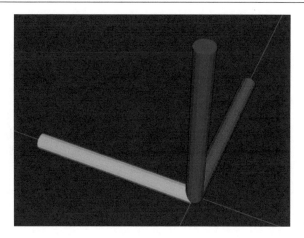

Figure 10.4 – Convention for coordinate systems in ROS

The arrows that you see between each frame in *Figure 10.3* are the relationship between each rigid part (link) of the robot. A TF is represented by an arrow.

> **Note**
> There can be some confusion between the names **links**, **frames**, and **TFs**. Let's make things clear:
> - Link: one rigid part of a robot
> - Frame: the origin of a link (axis in RViz)
> - TF: the relationship between two frames (arrow in RViz)

So, each rigid part will be linked to another rigid part, thanks to a TF. This transformation defines how those two parts are placed relative to each other. In addition to that, the TF also defines whether the two parts are moving, and if so, how—translation, rotation, and so on.

To make some parts of the robot move, you can move some of the cursors in the **Joint State Publisher** window. You will see the frames and TFs moving in RViz. If you also check the **RobotModel** box again, you will see the rigid parts moving as well.

To better understand, here is an analogy with the human arm: we can define the parts of the arm as *arm* (shoulder to elbow) and *forearm* (after the elbow). Those two are rigid parts (here, links) and do not move on their own. Each link has an origin frame, and there is a TF that defines where the arm and forearm are connected (imagine an axis in the elbow), and how they move (in this case, it's a rotation with a minimum and maximum angle).

As we will see later in this chapter, TFs are tremendously important. If the TFs for a robot are not correctly defined, then nothing will work.

Now you know what TFs are, but how are they all related to each other? As you can see on RViz, it seems that TFs are organized in a certain way. Let's go one step further and understand what the relationship between TFs is.

Relationship between TFs

In RViz, we have seen the links (rigid parts) and TFs (connections between the links). Links are mostly used for visual aspects in simulation and will be useful to define inertial and collision properties (when we work with Gazebo). TFs define how links are connected, and how they move between each other.

In addition to that, all the TFs for a robot are organized in a particular way, inside a tree. Let's explore the relationship between TFs and visualize the TF tree for the robot we started on RViz.

Parent and child

Each TF will be connected to another TF, with a parent/child relationship. To see one, you can, for example, disable all TFs on RViz, and only check the `base_link` and `gripper_pole` frames.

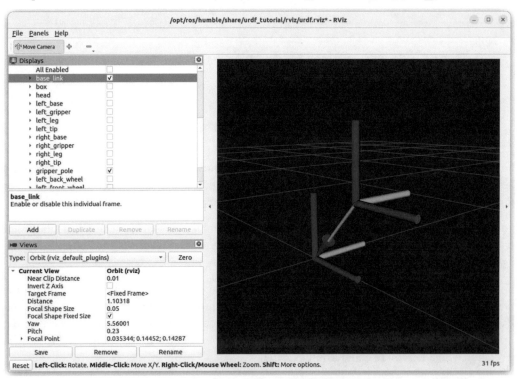

Figure 10.5 – The relationship between two frames

As you can see in this example, an arrow is going from the `gripper_pole` frame to the `base_link` frame. This means that `gripper_pole` is the child of `base_link` (or, `base_link` is the parent of `gripper_pole`).

If you look back at *Figure 10.3*, you can see all the frames for the robot, with all the relationships between them (TFs).

The order of those relationships is very important. If you move `gripper_pole` relative to `base_link` (the `gripper_extension` cursor in the **Joint State Publisher** window), then anything that's attached to `gripper_pole` (meaning children of `gripper_pole`) will also move with it.

That makes sense: when you rotate your elbow, your forearm is moving, but also your wrist, hand, and fingers. They don't move relative to the forearm, but as they are attached to it, they move relative to the arm.

Now, you can visualize all the links and TFs on RViz, see the relationship between TFs, and also how they are related to each other. Let's go further with the `/tf` topic.

The /tf topic

At this point, you might think that what we did in *Part 2* of this book has nothing to do with what we are doing now. Well, everything we have seen here is still based on nodes, topics, and so on.

Let's list all nodes:

```
$ ros2 node list
/joint_state_publisher
/robot_state_publisher
/rviz
/transform_listener_impl_5a530d0a8740
```

You can see that RViz is actually started as a node (`rviz`). We also have the `joint_state_publisher` and `robot_state_publisher` nodes, which we will come back to in the following chapters of this book. Now, let's list all topics:

```
$ ros2 topic list
/joint_states
/parameter_events
/robot_description
/rosout
/tf
/tf_static
```

Discovering TFs with RViz

You can see that the nodes that were started are using topics to communicate with each other. In this topic list, we find the `/tf` topic. This topic will be there for any robot you create. The TFs you saw on RViz are actually the 3D visualization of this topic—meaning the `rviz` node is a subscriber to the `/tf` topic.

You can subscribe to the topic from the terminal with the following:

```
$ ros2 topic echo /tf
```

If you do so, you will receive lots of messages. Here is an extract:

```
transforms:
- header:
    stamp:
      sec: 1719581158
      nanosec: 318170246
    frame_id: base_link
  child_frame_id: gripper_pole
  transform:
    translation:
      x: 0.19
      y: 0.0
      z: 0.2
    rotation:
      x: 0.0
      y: 0.0
      z: 0.0
      w: 1.0
```

This extract matches what we previously saw on RViz. It represents the transformation between `base_link` and `gripper_pole`. Here are the important pieces of information we can get from this message:

- Timestamp for the TF
- Parent and child frame IDs
- The actual transformation, with a translation and a rotation

> **Note**
>
> The rotation is not represented by Euler angles (x, y, z), but by a quaternion (x, y, z, w). Quaternions are usually better suited for computers, but it's difficult for humans to visualize them. Do not worry about this—we won't really have to deal with quaternions. If you ever have to do so in the future, you will have access to libraries that can convert angles into something you can understand.

One important thing we can get here is that the transformation is for a specific time. It means that with the topic data, you can follow the evolution of all TFs over time. You can know where `gripper_pole` is relative to `base_link`, now or in the past.

This `/tf` topic contains all the information we need, but it's not really human-readable. That's why we started with RViz, so we could see a 3D view with all the TFs.

Let's now finish this section by printing the TF tree, so we can see all the relationships in one single image.

Visualizing the TF tree

For each robot, you can visualize the complete TF tree in a simplified way, so that you can see all the relationships between all the TFs.

To do that, you will need to use the `tf2_tools` package. Make sure it is installed:

```
$ sudo apt install ros-<distro>-tf2-tools
```

Don't forget to source the environment after installing the package. Now, keep the robot running on RViz, and execute this command in a second terminal:

```
$ ros2 run tf2_tools view_frames
```

As you will see with the logs, it will listen to the `/tf` topic for five seconds. After this, the command exits with a big log that you can ignore.

You will get two new files, in the same directory as where you ran the command.

Open the PDF file. You will see something like this (I'm just adding the left side of the image, otherwise the text would be too small to read in the book):

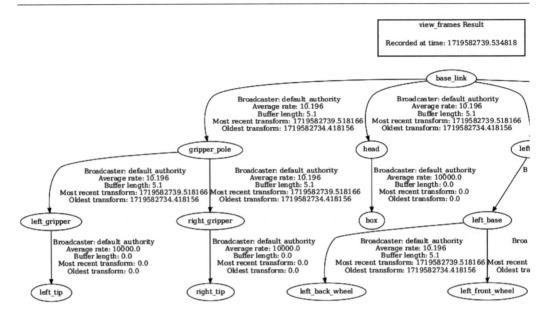

Figure 10.6 – TF tree for a robot

In this file, you get all the links and TFs at once, and it's quite clear to see which link is a child of which other link. Each arrow on the PDF represents a TF between the links.

As you can see, the root link for that robot is named `base_link` (the name `base_link` is used for most robots as the first link). This link has four children: `gripper_pole`, head, `left_leg`, and `right_leg`. Then, those links also get more children. Here, we can clearly see all the children for the `gripper_pole` link.

We can now understand that when we previously moved `gripper_pole` relative to `base_link`, then all children of `gripper_pole` were also moved relative to `base_link`.

> **Note**
> In ROS, one link can have several children, but only one parent. We will come back to this in the next chapter when we define the links and TFs ourselves.

In this example, we have just one robot. If you were to have several robots in your application, then you would have, for example, a `world` frame as the root link. Then, this frame would have several children: `base_link1`, `base_link2`, and so on. Each robot base link would be connected to the `world` frame. Thus, you can get a complete TF tree, not just for one robot but also for a complete robotics application with several robots.

Now, you have seen pretty much everything about TFs: what they are, how they are related to each other, and how they are organized. Let's finish this chapter by understanding what problem we are trying to solve with TFs.

What problem are we trying to solve with TFs?

You have now seen what TFs are and how you can visualize them for any ROS robot. This is great, but now we come to the final question for this chapter: why do we need to care about this? What problem are we trying to solve?

What we want to achieve

For a robotics application to work, we want to keep track of each 3D coordinate frame over time. We need a structured tree for all the frames of the robot (or robots).

There are two components here: we need to know where things are and when the transformations happened. If you remember, when we checked the /tf topic, you could see that for each parent and child frame, we had a transformation (translation and rotation in 3D space), and we also had a timestamp.

Here are some concrete examples of some questions you could need to answer in a robotics application:

- For one mobile robot, where is the right wheel relative to the left wheel and to the base of the robot? How does the wheel movement evolve over time?
- If you have an application with a robotic arm and a camera, then where is the camera relative to the base of the robot? And to the hand of the robot, so that the arm can correctly pick up and place objects that were detected by the camera?
- In another application with several mobile robots, where is each robot relative to the other ones?
- If you combine the two previous examples, where is the hand of the robotic arm relative to the base of one of the mobile robots?

So, with TFs, we want to know the following:

- How frames are placed relative to each other
- How they move relative to each other and over time

This is required for a robot to correctly work with ROS.

Now, let's see how you would compute the TFs by yourself, and how ROS automatically does this for you so you don't need to worry about it.

How to compute TFs

What is a transformation exactly? A transformation is the combination of a translation and a rotation in space.

As we are working in 3D, we have three components for the translation (x, y, z), and three components for the rotation (x, y, z). To find a transformation between two frames, you will need to compute those six elements, using 3x3 matrices.

I won't go into any mathematical details here, but you can probably guess that it won't be an easy task. Also, you need to compute this transformation for each frame, relative to the other frame. This increases the complexity.

For example, let's say you need to know where `left_front_wheel` is relative to `base_link`. Following the previous TF tree (open the PDF again), you can see that you need to follow this order: `base_link`, `left_leg`, `left_base`, and `left_front_wheel`.

Let's visualize that on RViz:

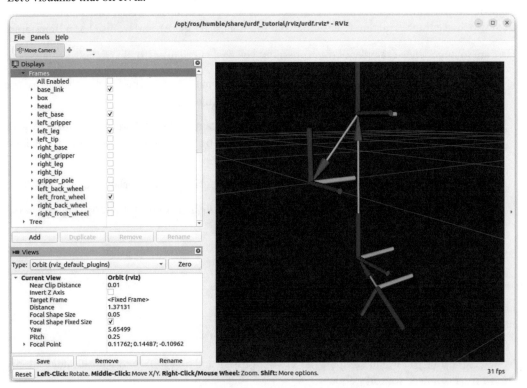

Figure 10.7 – Three transformations between four frames

You will need to compute three transformations in a row so that you can get this `base_link` to `left_front_wheel` transformation. You will have to repeat this for each frame relative to all other frames (the complexity then increases a lot as you add more frames), and track these over time.

This sounds like a lot of work. Fortunately, we don't have to do any of that, thanks to the ROS TF functionality. There is a library called `tf2`, and it already does that for us.

In the end, the biggest challenge with TFs is to understand how they work. You will mostly not use TFs directly in your application. Several packages will handle that for you. The only thing we need to do is to provide a robot description that specifies all the links and TFs for the robot. Then, using a package named `robot_state_publisher`, ROS will automatically publish TFs for us. That's what we will focus on in the next two chapters.

Summary

In this chapter, we started something a bit different. ROS is not all about programming; there are many additional things that make it a great tool for robotics.

You first discovered a 3D visualization tool used for ROS, named RViz. You will use this tool in most of your ROS applications. With RViz, you can visualize a robot model, which will be helpful when developing the model by yourself.

Then, you discovered what TFs are, and why they are so important in a ROS application. Here's a quick recap:

- We need to keep track of each 3D coordinate frame over time, for the entire robotics application (one or several robots).
- Instead of computing the transformations ourselves, we use the ROS TF functionality, with the `tf2` library. TFs are published on the `/tf` topic.
- TFs are organized into a structured tree that you can visualize.
- A TF defines how two frames are connected, and how they move relative to each other, over time.

To specify TFs for a robot, we will have to create a robot model, named *URDF*. This robot model will then be used by the `robot_state_publisher` node (we will see this a bit later) to publish the TFs. The published TFs will then be used by other packages in your application.

In the end, we won't really interact with TFs directly. The most important thing in this chapter is to understand what TFs are, and why we need them. This will help you understand what you're doing in the next chapter, when you create a robot model.

If things are still a bit confusing for now, don't worry too much. Continue with the next few chapters, and then come back to this TF chapter again and everything will make more sense.

Now, let's jump to the next chapter and create our first robot model.

11
Creating a URDF for a Robot

In the previous chapter, we started with an intuitive introduction to TFs, or TransForms. You have seen that TFs are very important; they will be the backbone of almost any ROS application. We concluded by saying that in order to generate TFs for a robot, you need to create a **Unified Robot Description Format** (**URDF**) file.

Basically, a URDF file will contain a description of all the elements of a robot. You will define each **link** (rigid part) of the robot. Then, to create relationships between the links, you will add some **joints**, which will be used to generate the TFs.

To write the content of a URDF, we will use XML. As you develop the URDF, you will be able to visualize it with RViz. This will be very helpful to see whether the links and joints/TFs are correct. We will also improve the URDF file and make it more dynamic with an additional tool named **Xacro**.

So, in this chapter, we are going to start the project for *Part 3*, with the URDF for a robot. We will create a mobile base with two wheels. This will be the foundation for the next few chapters. You can find the finalized URDF files inside the `ch11` folder of the book's GitHub repository (`https://github.com/PacktPublishing/ROS-2-from-Scratch`).

The hardest part about URDF is understanding how to assemble two links of a robot with a joint. Getting to do this without guidance is quite difficult because there are so many parameters and origins you can modify. I will explain the complete process, step by step, to make sure that you build something that works properly.

By the end of this chapter, you will be able to create your own URDF for almost any robot powered by ROS.

In this chapter, we will cover the following topics:

- Creating a URDF with a link
- The process of assembling links and joints
- Creating a URDF for a mobile robot
- Improving the URDF with Xacro

Creating a URDF with a link

In this section, you will dive directly in and create your first URDF. We will first create an XML file for the URDF. In this file, we will add a link, which will represent one rigid part of a robot, and visualize it in RViz. We will also explore the different types of shapes you can use—boxes, cylinders, and so on.

This will be a good first step, so you can get familiar with URDF and be ready to dive into the process of adding several links and joints together (in the next section).

Let's get started by setting up our URDF file.

Setting up a URDF file

A URDF file is simply an XML file with the `.urdf` extension.

Now, to keep things simple for this chapter, we will create a URDF file inside our home directory. In the next chapter, you will learn how to correctly package the URDF inside a ROS 2 application.

How should you name the URDF file? You could choose any name; it doesn't really matter. Usually, you will give it the name of your robot. If your robot's name is `abc`, then you will create an `abc.urdf` file. Let's use the name `my_robot`, as we previously did in this book.

Open a terminal and create a new file in your home directory:

```
$ cd
$ touch my_robot.urdf
```

You can then open this file with any text editor or IDE, for example, with VS Code:

```
$ code my_robot.urdf
```

Here is the minimum code you have to write inside a URDF file:

```
<?xml version="1.0"?>
<robot name="my_robot">
</robot>
```

We first open the file with the `<?xml version="1.0"?>` line, to specify that this file is an XML file—we also give the XML version.

Then, you need to open and close a `<robot>` tag. Everything you write in your URDF will be inside this tag. You also have to provide the name of the robot with the `name` argument.

Now, this is the minimum code for a URDF, but it will be useless if you don't define at least one element. Let's add a link inside this URDF.

Creating a link

You will now write your first link, which corresponds to one rigid part of a robot, and visualize it in RViz. With this, you will be able to see whether the link is correctly defined and modify it if necessary.

Let's start with the XML code to add the link.

Basic code for a link

To create a visual element for a link, you can use existing shapes: boxes, cylinders, and spheres (we will also see later how to include a custom shape made from **Computer-Aided Design** (**CAD**) software).

To get started, let's imagine the main base of a robot, represented as a box. The box is 60 cm x 40 cm x 20 cm, or `0.6 m x 0.4 m x 0.2 m`.

> **Note**
>
> In ROS, only the metric system is used. Here are a few units that we will use in this chapter:
>
> - Meters will be used for distances. If you have to specify 70 mm for example, you will write `0.07`.
>
> - Radians will be used for angles. 180 degrees corresponds to pi (about `3.14`) radians.
>
> - Meters per second will be used for velocity instances.

Here is the code for this first link:

```xml
<robot name="my_robot">
    <link name="base_link">
        <visual>
            <geometry>
                <box size="0.6 0.4 0.2" />
            </geometry>
            <origin xyz="0 0 0" rpy="0 0 0" />
        </visual>
    </link>
</robot>
```

Make sure to define the `<link>` inside the `<robot>` tag. Also, although indentation is not a requirement in XML, the best practice is to add some indentation to get a more readable file. Here, I used four spaces for each indentation.

Let's analyze the elements of this link. The `<link>` tag defines the link. All the properties for this link must be inside the tag. You also have to provide a `name` attribute for the link. As a convention, for the first link, we use `base_link`.

Then, inside this tag, we have the `<visual>` tag. If you want to define a visual appearance for the link (rigid part), you can do so with this tag. Inside, you will have the following:

- `<geometry>`: This will define the shape of the link. Here, we use the `<box>` tag and provide the dimensions with the `size` attribute.
- `<origin>`: This tag is quite important, as it defines the origin of the visual relative to the origin of the link. We will come back to this later in this chapter and see how to avoid confusion. The origin contains six elements for translation and rotation.

> **Note**
> The origin of rotation is written as `rpy`. This means *roll*, *pitch*, *yaw*. It's the same as x, y, and z, but using different names. Roll, pitch, and yaw are quite frequently used for aviation. You just need to get used to it for URDF.

As you can see, we first set all origins to 0. For now, the only thing we have specified are the dimensions of the box.

Visualizing the URDF in RViz

We have enough code to visualize the URDF in RViz. The goal here is to see the box in 3D and verify that everything is correct.

Doing this is very important when you develop a URDF. I would recommend always doing the following:

1. Make the smallest modification (add or modify something)
2. Visualize the URDF in RViz
3. If correct, continue with the next feature; if not, go back, fix it, and check again

Now, how can we visualize the URDF in RViz?

Good news: we can reuse the `urdf_tutorial` package (installed in *Chapter 10*) and provide our own URDF instead of the example ones. This is great because we can easily test a URDF file outside of a ROS 2 workspace, and we don't need to create any package yet.

Open a terminal and start the `display.launch.py` launch file, with the absolute path to your URDF file for the `model` argument:

```
$ ros2 launch urdf_tutorial display.launch.py \
model:=/home/<user>/my_robot.urdf
```

You will then see a box (red color by default) inside **RViz**. You will also have the **Joint State Publisher** window, empty, with no cursor.

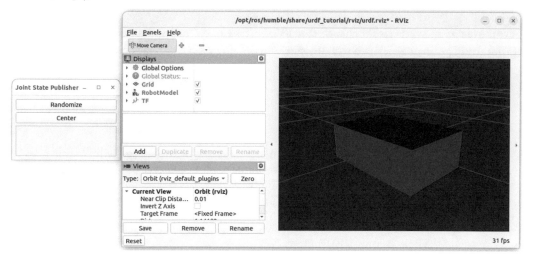

Figure 11.1 – Visualization of your URDF in RViz

If you go to **RobotModel** | **Links**, you will see the `base_link`. This is the link you have created; you can enable or disable the visual for that link.

Navigate around the box in **RViz**. You will see that the visual (box) is centered around the link origin. You can leave it like this or decide to offset the visual relative to the frame. Let's do this.

Modifying the origin of the visual

The link we have created is perfectly fine. However, we will offset the visual a bit so that the origin is not centered in the middle of the box, but instead, at the bottom of the box.

You don't necessarily need to do this. Sometimes, leaving the visual centered on the link origin is what you need. We will see some examples to illustrate this when we create the URDF for the mobile base a bit later in this chapter. For now, let's just assume we want to offset the visual.

To offset the visual, we will need to modify the `<origin>` tag inside the `<visual>` tag. In this `<origin>` tag, we have six elements for translation and rotation. We just want to move the visual up, so the only component we need to modify is the translation on the Z-axis (if you remember, with the right-hand rule, Z points up).

How much of an offset should we apply? As the box is currently centered on the link origin, we need to lift it up by half of its height.

We have defined the height as `0.2` m, so we need to offset the visual by `0.1` m.

Modify this line so that the z offset is `0.1`:

```
<origin xyz="0 0 0.1" rpy="0 0 0" />
```

Save the file, and to visualize the change, stop RViz (press *Ctrl* + *C* on the terminal where you launched it with `urdf_tutorial`), and start it again. You can do this every time you modify the URDF.

Now, you should see the box sitting on the ground, which means that the offset for the visual was correctly applied.

> **Note**
> The link origin is still the same; you only changed the visual relative to the link. This is an important distinction. If you feel confused, continue reading and everything will make sense after you see the full process with links and joints.

You have created your first link. Let's now see what kind of shapes you can use, and what customization you can add to the links in your URDF.

Customizing the link visual

A link is one rigid part of your robot. You could make it look like anything you want.

Let's explore the different shapes you can give to your links, and how to change their color.

Different shapes for a link

As you saw with the first link you created, you will define the shape of the link inside the `<geometry>` tag sitting in the `<visual>` tag.

There are three types of basic shapes you can use. For each one you will need to provide the dimensions with different attributes:

- `<box>`: You need to add a `size` argument with three components: x, y, and z
- `<cylinder>`: You need to add two arguments, `radius` and `length`
- `<sphere>`: You need only one argument, `radius`

We have just seen how to create a box in the previous code example. Here is an example of a cylinder of radius `0.2` m and length `0.5` m:

```
<geometry>
    <cylinder radius="0.2" length="0.5"/>
</geometry>
```

And an example for a sphere of radius `0.35` m:

```
<geometry>
    <sphere radius="0.35"/>
</geometry>
```

On top of those basic shapes, you can also use custom meshes that you export from CAD software, such as SolidWorks, Blender, and so on. You can use STL and Collada files, with `.stl` and `.dae` extensions, respectively. Setting up those files is not complicated but requires you to properly package your application around the URDF, which is something we will see in *Chapter 12*.

> **Note**
> There are even some tools that allow you to generate the complete URDF (links, joints, meshes, and so on) for a robot, directly from the CAD software. Great, isn't it? However, those tools are not always up to date or stable, and if you have an error, you could spend lots of time finding and fixing it. I recommend you write the URDF yourself and add the meshes one by one. You will end up with more control over what you're doing, and fixing errors will take much less time.

With the three basic shapes (box, cylinder, sphere), you can already do quite a lot and design a complete robot. We don't need more than that to get started, and we'll use them for the mobile robot that we'll create in this chapter. The link visuals have no effect on the TF generation, so this is not going to be a problem. Even when you start designing your own custom robot, you can start with basic shapes, and everything will work fine.

Here is the complete reference with every tag and attribute you can add in a link:

`https://wiki.ros.org/urdf/XML/link`

Let's now finish this section and see how to change the color of the link visual.

Link color

If you look at the first link on RViz, you can see that the visual color is red. This will be the default color for any basic shape you create.

As we add more shapes and combine them, it could be nice to modify their color, so that we can have some contrast between the different links. Otherwise, it will be hard to differentiate them on the screen.

To add a color to a link, you first need to create a `<material>` tag with a name. Then, you can use the color in your link visual.

Here is the complete code to make the link green:

```xml
<?xml version="1.0"?>
<robot name="my_robot">
    <material name="green">
        <color rgba="0 0.6 0 1" />
    </material>
    <link name="base_link">
        <visual>
            <geometry>
                <box size="0.6 0.4 0.2" />
            </geometry>
            <origin xyz="0 0 0.1" rpy="0 0 0" />
            <material name="green" />
        </visual>
    </link>
</robot>
```

Make sure the definition of the `<material>` tag is inside the `<robot>` tag, but outside of any `<link>` tag. In this new tag, you need to do the following:

- Define a name with the `name` attribute.
- Define a color with the `<color>` tag and `rgba` attribute (red, green, blue, alpha). Each of the four values should be between `0` and `1`. To create a basic, not-so-bright green color, we set the red and blue to `0`, and green to `0.6`. You can keep the alpha (transparency) set to `1`.

You only need to define this tag once, and then you can use it in any `<visual>` tag, inside any link. The color will apply to the basic shape. It should also apply to STL files if you have imported custom meshes (Collada files already contain the color, so there's no need for a `<material>` tag).

> **Note**
> When using the `<material>` tag inside a link, make sure to place it inside the `<visual>` tag, but not inside `<geometry>`. The `<geometry>`, `<origin>`, and `<material>` tags should be direct children of the `<visual>` tag.

Great, you can now create a link with different kinds of shapes and colors. This is a great start so you can represent any rigid part of your robot in 3D.

Let's now see how to assemble different links together and thus create a complete robot model.

The process of assembling links and joints

Now that you have a URDF file with one link, let's add another link, and connect them with a joint. This joint will be used to generate a TF.

Properly assembling two links with a joint is the main problem anybody faces when learning URDF. There are several origins and axes you can modify, and getting two parts to be correctly placed between each other, with the correct movement, can be challenging.

In this section, we will focus on that process, so it becomes easier for you. I have condensed it into five steps that you can follow in the order every time you add a new link.

After you get confident with the process, you will be able to create a URDF for any kind of robot. A complete robot model is just a sequence of links connected to each other. If you correctly understand how to do it for two links, adding twenty more links won't be that difficult.

We will also explore the different kinds of joints you can use in a URDF. To verify that everything works fine, we will use RViz, as well as the TF tree generated by the `tf2_tools` package.

This section is very important, and I recommend that you do not skip it. Also, feel free to come back to it later, whenever you have some doubts about how to connect two links.

Let's get started with the first step of the process: adding a second link to the URDF.

Step 1 – adding a second link

For this example, we want to add a cylinder (radius: `0.1` m, length: `0.3` m) on top of the box. We will make this cylinder gray (to create a contrast with the green box), so, let's first create another `<material>` tag with gray color.

You can add this new tag just after the previous `<material>` tag:

```
<material name="gray">
    <color rgba="0.7 0.7 0.7 1" />
</material>
```

Then, let's add the link. Create another `<link>` tag inside the `<robot>` tag, with the specifications for the second link. You can place this link just after the `base_link`:

```
<link name="shoulder_link">
    <visual>
        <geometry>
            <cylinder radius="0.1" length="0.3" />
        </geometry>
        <origin xyz="0 0 0" rpy="0 0 0" />
        <material name="gray" />
```

```
        </visual>
    </link>
```

You can give any name for the link. Here, I specify `shoulder_link` because we will create the beginning of a robotic arm for this example. You could then have several parts: base, shoulder, arm, forearm, hand, and so on. The best practice is to give meaningful names to the links of your robot.

As you can see, we set all the origin elements to 0. This is quite important and will be the first step of the process: you add a link, but you don't modify the origin.

Now, if you try to visualize the URDF in RViz (stop and start from the terminal), you will get an error. Among the logs, you will see this:

```
Error: Failed to find root link: Two root links found: [base_link] and
[shoulder_link]
```

You get this because all links in a URDF need to be related to each other with a parent/child relationship, as we saw in the previous chapter. Here, there is no explicit relationship, so ROS cannot know which one is the parent, and which one is the child.

We will define this relationship with a joint. This will also allow us to generate our first TF for the robot.

Step 2 – adding a joint

To define how two links are connected, you need to add a joint. Here is the code you can write after the two `<link>` tags (and still inside the `<robot>` tag):

```
<joint name="base_shoulder_joint" type="fixed">
    <parent link="base_link" />
    <child link="shoulder_link" />
    <origin xyz="0 0 0" rpy="0 0 0" />
</joint>
```

To create a joint, you add a `<joint>` tag, which contains two attributes:

- name: You can choose whatever you want, as long as it makes sense. I usually combine the names of the two links I want to connect: `base_link` and `shoulder_link` become `base_shoulder_joint`.
- type: We will come back to the different joint types you can use in *Step 4*. For now, we set it as `fixed`, which means that the two links won't move between each other.

Inside the `<joint>` tag, you then have three more tags:

- `<parent>`: This is the parent link. You have to write the exact name of the link with the `link` attribute.
- `<child>`: You will write the exact name of the child link with the `link` attribute.
- `<origin>`: This will define the origin of the child link relative to the origin of the parent link. Once again, we use `xyz` for translation and `rpy` for rotation.

This is the second step of the process: you add a joint between the two links and define which one is the parent and which one is the child. For now, you leave all origin elements to 0.

With this code, you can start RViz again, and this time, as there is a relationship between the links, the URDF will be displayed. Here is what you will get:

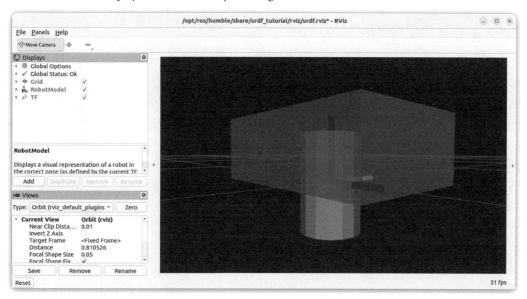

Figure 11.2 – Two links and one joint, with all origins set to 0

As you can see, we now have a box (`base_link`) and a cylinder (`shoulder_link`). As all origin elements are set to 0, both link origins are at the same place.

On top of that, you can also validate that the TF you've created with the joint is correctly placed in the TF tree. With the `ros2 run tf2_tools view_frames` command, you can generate the TF tree. In the newly created PDF file, you will see this:

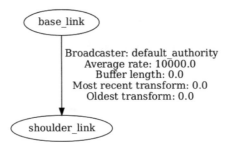

Figure 11.3 – TF tree to validate the relationship between the two links

We can validate that the relationship we have defined is correct. We now need to correctly place the `shoulder_link` relative to the `base_link`.

Step 3 – fixing the joint origin

This is the step where most people get confused. If you look at the current code, we have three `<origin>` tags: one in each link and one in the joint. So, which origin do we need to modify?

The classic error is to start to modify several random origins at once and try to find something that works by tinkering with the values. Even if that seems to work in the end, it will probably create more problems when you add other joints.

So, for this step, I emphasize that you have to follow the *exact* process I will describe, every time.

The first thing to do is to modify the `<origin>` tag of the joint, so you get the frame of the child link correctly placed. This is what matters the most. You will first fix the joint origin, and then, and *only* then, fix the visual origin.

For this, expand **RobotModel** and **Links**, and uncheck the `shoulder_link` visual (at this stage, seeing the visual can cause confusion, so we disable it). Then, ask yourself this question: where should be the frame for the `shoulder_link`, relative to the frame of the `base_link`?

We want the `shoulder_link` to be on top of the box, so we need to move the frame by the height of the box; here, that's `0.2` m. There is no rotation needed for this joint, just a translation.

So, you can now modify the `<origin>` tag inside the `<joint>` tag:

```
<origin xyz="0 0 0.2" rpy="0 0 0" />
```

Start RViz again, disable the visual for the `shoulder_link`, and check whether the frame is correctly placed.

The process of assembling links and joints   265

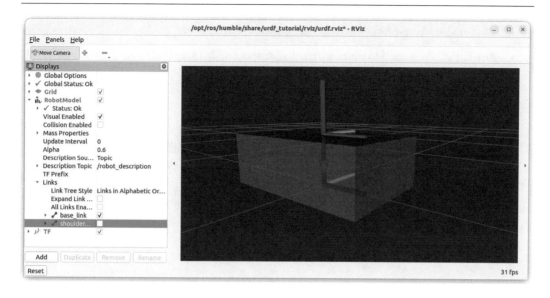

Figure 11.4 – Setting the joint origin without the visual

Great, it seems that the frame for the `shoulder_link` is in the right place: on top of the box.

This origin is the most important; this is what will define the TF. In the end, TFs will be generated with all the joint origins in your URDF. The visual will not be taken into account (visuals will be used for inertial and collision properties in the Gazebo simulator).

We have our joint origin correctly set up. Now, let's see how to specify the joint type, so we can define the movement between the two links.

Step 4 – setting up the joint type

To keep things simple for the previous explanation, we have set the joint type as *fixed*, which means that the two links are not moving relative to each other.

Lots of joints you will create will be like this. For example, if you place a sensor (camera, lidar) on your robot, the sensor won't move. You can create a link for the sensor, and then a fixed joint to connect this sensor to your robot.

However, for some rigid parts (an arm in a robotic arm, wheels, torso, and so on), you will need to specify that the child link is moving relative to its parent link. I will not describe all possible joint types, but the most common movements you will find in any robot will be as follows:

- **Fixed**: As previously mentioned, this is if you have two parts that are not moving
- **Revolute**: A rotation with a minimum and maximum angle, for example, in robotic arms

- **Continuous**: An infinite rotation, usually used for wheels
- **Prismatic**: If you need to make a part of your robot slide (only translation, no rotation)

You can find the complete reference for all joint types at http://wiki.ros.org/urdf/XML/joint. There, you can get all possible elements you can add in a `<joint>` tag. Most of them are optional.

Coming back to our example, let's say that the shoulder link is rotating (with a minimum and maximum) on the Z-axis on top of the box. You can specify this with the `revolute` joint type. By doing this, you will also need to add an `<axis>` tag to specify the axis of rotation, and a `<limit>` tag for the minimum and maximum angle. Let's modify the `<base_shoulder_joint>` tag:

```
<joint name="base_shoulder_joint" type="revolute">
    <parent link="base_link" />
    <child link="shoulder_link" />
    <origin xyz="0 0 0.2" rpy="0 0 0" />
    <axis xyz="0 0 1" />
    <limit lower="-3.14" upper="3.14" velocity="100" effort="100"/>
</joint>
```

When choosing `revolute`, we first need to define which axis will be rotating. As we chose z (if you look at *Figure 11.4*, we want to rotate around the blue axis), we write `"0 0 1"`, which means: no rotation on x and y, and a rotation on z.

We set the revolution between -180 and +180 degrees (about `-3.14` and `3.14` radians). We also have to specify a value for the velocity and effort limits. Those two will usually be overridden by other ROS nodes. Set them to `100` by default; it won't be important here.

With this joint, you have created a TF that defines the position of the `shoulder_link` relative to the `base_link`, and the movement between those two links.

Now, you can start RViz again (disable the visual), and you will find a cursor in the **Joint State Publisher** window. Move the cursor to make the `shoulder_link` rotate on top of the `base_link`. Once again, it's all about making the frame correctly move. If the visual still doesn't look right, don't worry about it; this is what we will fix in the last step of this process.

Step 5 – fixing the visual origin

We can now fix the origin of the `shoulder_link` visual. Don't modify the `base_link` origin, as it's already correct. Here, we only modify the visual for the child link.

To do so, you can enable the `shoulder_link` visual again on RViz and see that the frame is at the center of the cylinder. Thus, we need to offset the visual by half of the length of the cylinder, which means `0.15` m (half of `0.3` m).

Modify the `<origin>` tag inside the `<link>` tag of the `shoulder_link`:

```
<origin xyz="0 0 0.15" rpy="0 0 0" />
```

If you start RViz again, you will then see that everything is correctly placed:

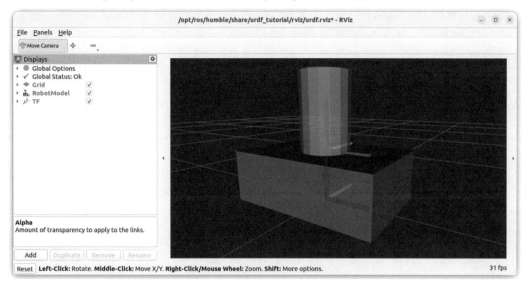

Figure 11.5 – The end of the process for fixing the origins

> **Note**
> In this example, we want the cylinder to be placed on top of the frame. If you wanted the frame to be in the center of the cylinder, you would place the joint origin higher on the Z-axis, and then leave the visual origin as it is (I talk more about this later in this chapter, when we work on the wheels of the mobile robot).

The process is now finished. As it is very important, let's now do a quick recap.

Recap – the process to follow every time

When you create a URDF, you will first start with a link, usually named `base_link`. Then, for each link that you add, you will also add a joint, to connect this new link to an existing one.

Here is the process to add a new link to your URDF:

1. Add a new `<link>` tag and set all origin elements to 0.
2. Add a new `<joint>` tag. You have to specify a parent and child link. The parent link will be an existing link you've already created, and the child link is the new link you've just added. Set all origin elements to 0.

3. Fix the origin of the joint. Visualize the URDF in RViz, disable the visual for the new link, and figure out where the frame for the new link should be relative to its parent.
4. If the joint is associated with a movement, set the joint type. Depending on the type, you might have to set the axis of rotation/translation, some limits, and so on.
5. Once the frame origin is correct, enable the visual in RViz and fix the visual origin for the link (only for the child link, not the parent link).

After this is done, congratulations, you have successfully connected two links, and the joint will be used to generate a TF. You can repeat this process for each new link that you add to your URDF.

A few important things to remember are as follows:

- Only add one link and one joint at a time. Finish the process, then add another link.
- Don't modify the origin of the parent link or any link you've already created before. This is a sure way to start messing up with the URDF and spending hours on debugging. If you need to go back to a previous link, then disable all children, fix the link, and continue from there.
- One link can have several children, but only one parent.
- After each modification, as small as it can be, always verify in RViz. Don't try to modify several origin elements at the same time. Change one, validate it, then go to the next one.
- You can verify that the relationship between all links is correct by printing the TF tree.

In the end, this process is not that complicated. Following these steps to the letter will ensure that you build your URDF right the first time. Not only will you be sure about what you are doing but, it will save you lots of time in the long term, and your URDF will be cleaner.

We now have all the information we need to create a complete URDF for a robot.

Writing a URDF for a mobile robot

You have seen the complete process of adding links and joints in a URDF file. By repeating this process several times, you can create a complete model for any robot.

We are now going to use this process to create a mobile robot. We will then use this URDF as a base for the next chapters.

I will first show you the final robot model so you get an idea, and then we will build the URDF, step by step. You will get the specs for the robot as we build it. I encourage you to follow along and even write the code at the same time as you read the section. This is good practice for you to get more comfortable with URDF. As a reminder, you can find the complete code on GitHub.

What we want to achieve

Before writing any code, it's important to define what we want to achieve. The final result will look like this in RViz:

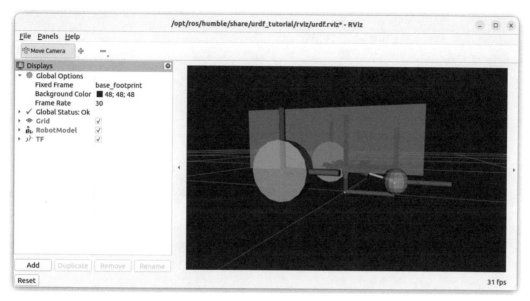

Figure 11.6 – Final result for the mobile robot

We will start the URDF with the robot's main structure (chassis), represented by a box. This box is the same as the one we have created with the `base_link`.

Then, we will add two wheels, on each side of the chassis. Those two wheels will have a continuous rotation. Finally, for stability, we add a caster wheel (sphere) that will help the robot not fall forward when we simulate it in Gazebo. This caster wheel will be a fixed joint, and we won't add any movement to it.

On top of that, at the end, we will also add another link named `base_footprint` (with no visual, we can consider it as a virtual link), which is the projection of the `base_link` on the ground. I will explain this further when we do it.

To get started, go back to the `my_robot.urdf` file we have written previously, and keep the `base_link`. Remove the `shoulder_link` as well as the `base_shoulder_joint`.

Let's now add the two wheels to the side of the robot.

Adding the wheels

We will add the wheels one by one, starting with the right wheel. Unless you're an expert already, it's important that you only add one link and one joint at a time.

Right wheel

To add the right wheel, we will follow the five-step process we just described. Let's name this link `right_wheel_link`. For the visual, we will use a cylinder, with a radius of 0.1 m and a length of 0.05 m. Here is the code for the link (*Step 1*):

```
<link name="right_wheel_link">
    <visual>
        <geometry>
            <cylinder radius="0.1" length="0.05" />
        </geometry>
        <origin xyz="0 0 0" rpy="0 0 0" />
        <material name="gray" />
    </visual>
</link>
```

As the wheel will be attached to the `base_link` (green color), we choose the gray color to make a contrast. As you can see, we set all origin elements to 0. We will only come back to those elements at the end of the process.

Now, let's add a joint between the base and the right wheel (*Step 2*).

```
<joint name="base_right_wheel_joint" type="fixed">
    <parent link="base_link" />
    <child link="right_wheel_link" />
    <origin xyz="0 0 0" rpy="0 0 0" />
</joint>
```

When you get more experienced, you can set the movement type when creating the joint, but let's go step by step. For now, we set the type to `fixed`, so we can set the joint origin first, and then specify the type of movement. This will make things simpler and less prone to errors.

With this code, you can already visualize the URDF in RViz. Disable the `right_wheel_link` visual (**RobotModel | Links**) as we don't need this for now, and it could make us confused when setting the joint origin.

Then, the question is: where do we place the `right_wheel_link` frame relative to the `base_link` frame (*Step 3*)? Let's see for each axis:

- **X translation (red axis)**: We want the wheel to be a bit behind, let's choose a -0.15 m offset.
- **Y translation (green axis)**: The wheel should be on the side of the robot. Here, you have two choices (to understand this, make sure to have RViz open with the URDF):
  - You can place the wheel origin just on the right side of the box. As the *Y*-axis is pointing left, the offset would be -0.2 m (half of the width of the box). Then, later on, you would need to add an offset in the wheel visual, like we did for the `shoulder_link` in the previous example.

- Alternatively, you can add a small additional offset to the joint, so that the wheel would be outside of the box, and the visual would be centered around the frame. This second option is a good idea for wheels and some sensors—for example, when using a lidar, this is required to make the scan work properly. We then need to add -0.2 m, and an additional -0.025 m (half of the length of the wheel). The total offset is -0.225m.

- **Z translation (blue axis)**: There is no need for any offset here, as we want the center of the wheel to be at the bottom of the box.

We have our three values for the translation. That's all we need. You might think that we have to rotate the joint axis because the visual is not correctly orientated. However, this is one of the most common mistakes, and it's where you could start to modify the wrong values to fix the wrong problem. As we saw in the process, we first fix the joint with the visual disabled, and then, and only then, we fix the visual.

Let's modify the `<origin>` tag inside the `<joint>` tag (again: not in the link, only in the joint):

```
<origin xyz="-0.15 -0.225 0" rpy="0 0 0" />
```

Then, start RViz again, disable the visual for the wheel (this visual is still wrong, but not a problem), and you can see that the joint is placed a bit outside the box.

We can now easily add the movement (*Step 4*). As the wheel will continuously rotate (there is no minimum or maximum position), we choose the `continuous` type. We also need to specify the rotation axis. By looking at the robot model in RViz, we can see that we have to pick the *Y*-axis (the wheel should rotate around the green axis).

Thus, we modify the `<joint>` tag accordingly:

```
<joint name="base_right_wheel_joint" type="continuous">
    <parent link="base_link" />
    <child link="right_wheel_link" />
    <origin xyz="-0.15 -0.225 0" rpy="0 0 0" />
    <axis xyz="0 1 0" />
</joint>
```

There is no need to specify a `<limit>` tag for `continuous`, as we did with the `revolute` type previously.

You can now start RViz again, disable the wheel visual, and move the new cursor named base_right_wheel_link on the **Joint State Publisher** window. You should see the joint correctly rotating around the base.

> **Note**
> With the cursor, you will see a minimum of about -3.14 and a maximum of about 3.14 (total of 360 degrees). Don't worry about this, it's just a graphical element. As the joint type is continuous, there will be no minimum or maximum position when we control it later on.

That's it for the joint. With this, the TF will be correctly generated. We can now finish the process and fix the visual for the link (*Step 5*). If you re-enable the wheel visual, you will see that it's not correctly orientated. You would need to add a 90-degree rotation on the *X*-axis (around the red axis). This corresponds to pi/2, or about `1.57` radian (we will see later in this chapter how to use a precise value of pi).

Let's modify the `<origin>` tag for the `right_wheel_link`:

```
<origin xyz="0 0 0" rpy="1.57 0 0" />
```

Start RViz again, and now everything should be fine: the wheel visual will be correctly placed (just outside of the box) and with the right orientation. When you move the cursor for the joint, the wheel will turn correctly.

The right wheel was probably the most complicated part of this robot, but as you can see, if you follow the process in the order, there should be no problem. You can be sure that all values are correct, and that there is no error that will propagate onto the next links that you add.

Let's now write the code for the left wheel.

Left wheel

As the left wheel is the same as the right wheel, but on the opposite side of the box, writing the code will be fairly straightforward. Here I won't repeat the full process and will just show you the final `<link>` and `<joint>` tags.

Let's start with the link:

```
<link name="left_wheel_link">
    <visual>
        <geometry>
            <cylinder radius="0.1" length="0.05" />
        </geometry>
        <origin xyz="0 0 0" rpy="1.57 0 0" />
        <material name="gray" />
    </visual>
</link>
```

This is the same code as for the `right_wheel_link`.

Then, the left wheel will be connected to the chassis, so the parent of the `left_wheel_link` will be `base_link`. Here is the code for the joint:

```
<joint name="base_left_wheel_joint" type="continuous">
    <parent link="base_link" />
    <child link="left_wheel_link" />
    <origin xyz="-0.15 0.225 0" rpy="0 0 0" />
```

```
    <axis xyz="0 1 0" />
</joint>
```

Everything is the same, except for the offset on the *Y*-axis. For the right wheel, we had to go on the negative side, but for the left wheel, we went on the positive side.

The important thing to check is that when you move the two cursors on the **Joint State Publisher** window on the positive side, both wheels are rotating in the same direction. If you have this, then your differential drive system is correctly designed.

We can now add the **caster wheel** for the stability of the robot.

Adding the caster wheel

When we add physics and gravity later on in the Gazebo simulator, you can guess that the robot would fall on the front side, because it's out of balance. To fix this, we will add what's called a caster wheel. It's the same principle as the wheels under a desk chair.

To make things simple, the caster wheel will be represented by a sphere (with a radius of 0.05 m). For the movement, even if the wheel is rotating, this is not a rotation we control with ROS. It is a passive rotation; thus, we will consider the joint as fixed.

Let's create the link first (*Step 1*):

```
<link name="caster_wheel_link">
    <visual>
        <geometry>
            <sphere radius="0.05" />
        </geometry>
        <origin xyz="0 0 0" rpy="0 0 0" />
        <material name="gray" />
    </visual>
</link>
```

There's nothing too complicated here. Now, the caster wheel will be connected to the chassis of the robot. Let's add the joint (*Step 2*). To follow the process, we first set all origin elements to 0:

```
<joint name="base_caster_wheel_joint" type="fixed">
    <parent link="base_link" />
    <child link="caster_wheel_link" />
    <origin xyz="0 0 0" rpy="0 0 0" />
</joint>
```

Start RViz and disable the visual. From this, let's see where to place the origin of the caster wheel relative to the origin of the base (*Step 3*):

- **X translation**: As the two wheels are on the back of the robot, the caster wheel should be at the front. We can choose for example `0.2` m.
- **Y translation**: For better stability, we want the caster wheel to be centered, so, `0`.
- **Z translation**: I specified `0.05` as the radius so that the diameter of the caster wheel (0.1 m) corresponds to the radius of the wheel. Thus, in order for the wheels and caster wheels to be aligned on the ground, we need to offset the Z-axis by `-0.05` m. If you are not so sure about this, it's simple: try some values and see the results in RViz.

There is no need to set any rotation for the joint, as we won't move the caster wheel, and it's a sphere. Let's apply the offset in the `<origin>` tag of the `<joint>` tag:

```
<origin xyz="0.2 0 -0.05" rpy="0 0 0" />
```

Now the caster wheel will be correctly placed under the chassis, and you can verify that the bottoms of both wheels and caster wheel seem to be aligned. Here, there is no need to set a movement type (*Step 4*) nor to fix the visual (*Step 5*).

The robot model is now finished. There is just one more thing we will do to better prepare the robot for the following.

Extra link – base footprint

The robot is correctly designed and will work fine when we add control to it. However, there is one improvement we can make.

If you look at the robot in RViz, the `base_link` frame is not at the same altitude (z offset) as the bottom of the three wheels. This is fine, but it would be nice to have the origin of the robot aligned with the ground where the robot will be.

Not only will it make the robot look better in RViz (although it's not important), but it will also make things easier in the future. An example would be when you want to create a transformation from a docking station to a mobile robot, or from one robot to another one. Another example would be if you want to use the **Navigation 2** stack. If all robot origins are on the ground, you can then work in 2D, which is easier to handle.

For that reason, it's quite common to add a virtual link named `base_footprint`, which will be the projection of the `base_link` on the ground. We say the link is *virtual* because it doesn't contain any visuals; it's just an additional frame we define in the space. Here is the code for the link:

```
<link name="base_footprint" />
```

As you can see, this link is extremely simple, as we didn't include any `<visual>` tag. For a link name, we usually start with the rigid part name and add the `_link` suffix. Here, we make an exception. You will find this `base_footprint` name in many URDF files for existing mobile robots.

Now, we can add a new `fixed` joint, with the `base_footprint` as the parent, and the `base_link` as the child:

```
<joint name="base_joint" type="fixed">
    <parent link="base_footprint" />
    <child link="base_link" />
    <origin xyz="0 0 0.1" rpy="0 0 0" />
</joint>
```

We apply a `0.1` m offset on the Z-axis, which corresponds to the right and left wheels' radius.

> **Note**
>
> To organize all links and joints in a URDF, I usually write all the links first, followed by all the joints. You could also decide to alternate between links and joints. There is no right or wrong method; it's a question of preference.

You can now visualize the final result on RViz. To get the correct view, click on **Global Options**, and in the **Fixed Frame** menu, choose `base_footprint`. You will see that the bottoms of the wheels, and the `base_footprint`, are aligned on the ground.

While RViz is still running, you can print and visualize the TF tree:

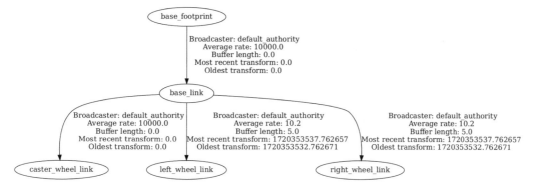

Figure 11.7 – Final TF tree for the mobile robot

We are now done with the URDF. Before we finish this chapter, let's explore Xacro, which will allow you to improve your URDF file and make it more scalable.

Improving the URDF with Xacro

The more complex your robot, the bigger the URDF. As you add more links and joints, you will end up having problems scaling your robot model. Also, what we have written so far is not so dynamic: all the values are hardcoded.

Xacro is an additional ROS feature you can use to solve all those issues. We will now see how to make a URDF file compatible with Xacro, how to create variables and functions, and how to split your URDF into several files.

With Xacro, your URDF files will become more dynamic and scalable. All serious ROS 2 projects use Xacro, so it's important to learn how to work with it.

Let's get started with the setup.

Making a URDF file compatible with Xacro

We will start by making sure our URDF file can use Xacro features. Before doing anything, let's make sure that Xacro is installed (it should already be there with all the previous packages we installed):

```
$ sudo apt install ros-<distro>-xacro
```

Now, to use Xacro in your URDF file, you need to make two changes.

First, change the file extension. The file is currently named `my_robot.urdf`. For Xacro, you will use the `.xacro` extension. A common practice is to use the `.urdf.xacro` extension for the main URDF file of your robot, so the file would be named `my_robot.urdf.xacro` (the important thing is to have `.xacro` at the end).

Once you've changed the extension, open the file and modify the `<robot>` tag:

```
<robot name="my_robot" xmlns:xacro="http://www.ros.org/wiki/xacro">
```

Every time you want to use Xacro in a URDF file, you will have to add this `xmlns:xacro` argument.

That's it for the setup. Now, to visualize the URDF in RViz, you will run the same command as before, with the new file name:

```
$ ros2 launch urdf_tutorial display.launch.py model:=/home/<user>/my_robot.urdf.xacro
```

Let's now discover how to use variables with Xacro.

Xacro properties

In programming, one of the first and most important things you learn is how to use variables. Variables do not exist in URDF, but you can use them with Xacro. Here, a variable is called a **property**.

With Xacro properties, we will be able to specify values such as dimensions at the beginning of the file and use those properties inside the `<link>` and `<joint>` tags. This way, values and computations are not hardcoded. It will make things less confusing and more readable. Plus, if we ever need to modify a dimension of the robot, we just need to modify one value at the beginning of the file.

Also, it's important to note that Xacro properties are considered constant variables. After you set their value, you won't modify them anymore.

To declare and define a Xacro property, you will use the `<xacro:property>` tag and provide two arguments: `name` and `value`.

Let's declare a few properties at the beginning of the file (inside the `<robot>` tag):

```
<xacro:property name="base_length" value="0.6" />
<xacro:property name="base_width" value="0.4" />
<xacro:property name="base_height" value="0.2" />
<xacro:property name="wheel_radius" value="0.1" />
<xacro:property name="wheel_length" value="0.05" />
```

Those are all the values we need for computing everything else in the URDF.

Then, to use a Xacro property, you simply have to write `${property_name}`. You can also do computations. For example, to multiply a value by 2.5, you will write `${property_name * 2.5}`. With this information, let's modify the content inside the `base_link` to remove any hardcoded value:

```
<geometry>
    <box size="${base_length} ${base_width} ${base_height}" />
</geometry>
<origin xyz="0 0 ${base_height / 2.0}" rpy="0 0 0" />
```

As you can see, we specify the box size with only properties. The most interesting part is how we compute the visual offset on the Z-axis. Writing `${base_height / 2.0}` is much more explicit than just writing `0.1`. Not only is it more dynamic, but we also have a better idea of what this computation is about. Imagine coming back to this URDF in six months and trying to figure out why the offset value is `0.1`, without any context. With this property, there is no possible doubt.

Let's now modify the visual for the links of the right wheel and left wheel:

```
<geometry>
    <cylinder
        radius="${wheel_radius}" length="${wheel_length}"
    />
</geometry>
<origin xyz="0 0 0" rpy="${pi / 2.0} 0 0" />
```

> **Note**
> The Xacro property, or constant pi, is already defined. Instead of hardcoding an approximation of pi, you can just use it with `${pi}`.

Finally, here is the code for the caster wheel:

```
<geometry>
    <sphere radius="${wheel_radius / 2.0}" />
</geometry>
<origin xyz="0 0 0" rpy="0 0 0" />
```

As you can see, we didn't define a `caster_wheel_radius` property. This is because the caster wheel radius needs to be proportional to the right and left wheel radius. It needs to be half of the value so that both wheels and caster wheels can touch the ground while making the robot stable. By using a Xacro property here, if we were to modify the wheel radius, then the caster wheel would automatically resize to make the robot stable.

We have now modified all the links; let's change the values in the joint origins. For the `base_joint`, we have the following:

```
<origin xyz="0 0 ${wheel_radius}" rpy="0 0 0" />
```

The `base_right_wheel_joint` is a bit more complex. However, once again, by writing this, we will make the computation more readable and less prone to errors in the future:

```
<origin xyz="${-base_length / 4.0} ${-(base_width + wheel_length) / 2.0} 0" rpy="0 0 0" />
```

The `base_left_wheel_joint` will be the same, except that the sign on the Y-axis is positive. We finish with the `base_caster_wheel_joint`:

```
<origin xyz="${base_length / 3.0} 0 ${-wheel_radius / 2.0}" rpy="0 0 0" />
```

All those changes in the code didn't modify the robot model. When visualizing it in Rviz, it should look the same.

To test that everything worked, try modifying the values for some Xacro properties at the beginning of the file. The robot model in RViz will have different dimensions, but it should still make sense.

That's it for Xacro properties. This concept is not that hard to understand and apply, especially if you're already familiar with variables and constants. Let's now switch to functions, or macros.

Xacro macros

A Xacro macro is the equivalent of a function in programming. With Xacro, a macro works like a template: it's a piece of XML code that you can reuse with different values (parameters). A macro doesn't return anything.

Macros are quite useful when you have to duplicate a link or a joint several times. Imagine a robot with four cameras. You can create a macro for the camera link, and then call the macro instead of re-writing the same code four times.

With our robot model, we have almost the same code for the right_wheel_link and the left_wheel_link. The only difference is the name of the link. Let's create a macro for those links.

To create a macro, you will use the <xacro:macro> tag and give a name as well as a list of params. You can specify zero, or as many parameters as you want—just separate them with a space. Here is an example:

```
<xacro:macro name="wheel_link" params="prefix">
    <link name="${prefix}_wheel_link">
        <visual>
            <geometry>
                <cylinder radius="${wheel_radius}"
                          length="${wheel_length}" />
            </geometry>
            <origin xyz="0 0 0" rpy="${pi / 2.0} 0 0" />
            <material name="gray" />
        </visual>
    </link>
</xacro:macro>
```

This piece of code won't do anything by itself. We need to call it, just like you would call a function. To call a macro, you will write the following:

```
<xacro:name param1="value" param2="value" />.
```

Remove the right_wheel_link and the left_wheel_link, and write this instead:

```
<xacro:wheel_link prefix="right" />
<xacro:wheel_link prefix="left" />
```

The "right" value in the prefix parameter will be applied to the link name inside the macro, making the name right_wheel_link. The same thing applies for the left wheel.

As you can see, macros can help you reduce code duplication. In this example, the benefit is not as big, but if you need to duplicate some links or joints more than three times, then macros can be extremely useful. Also, if you are creating part of a URDF that will be used by other people, writing a macro can help them to easily integrate your code into theirs, and customize it with the different parameters they can give as input.

Xacro properties and macros will allow you to make one URDF file more dynamic. Let's finish this section by seeing how to make a URDF even more scalable.

Including a Xacro file in another file

Thanks to Xacro, you can split your URDF into several files.

This is useful to separate different parts of your robot, for example, the main core base, and extra sensors that you add on top. If you combine two robots, for example, a robotic arm on top of a mobile robot, you can create one URDF for each robot and combine them into a third one. Another benefit is collaboration. Creating a macro inside one file that other developers can include will make it easier for them to work with your code.

Coming back to our URDF, let's split the file into three:

- `common_properties.xacro`: This will contain the material tags and other properties that could apply to any part of our robotics application
- `mobile_base.xacro`: This file will contain the properties, macros, links, and joints that are specific to the mobile base
- `my_robot.urdf.xacro`: In this main file, we include the two previous files

By doing this, we will make the URDF more dynamic and easier to modify in the future. If you want to combine several mobile bases or add other robots or sensors, you can create more Xacro files that you include into the main one (`my_robot.urdf.xacro`).

Now, create the two additional files, and let's see how to organize the code. Make sure to put all three files in the same directory. In the next chapter, we will install them properly inside a ROS 2 package, but for now, keep them all in the same place (it will make it easier to include the first two files into the third one, using the relative path).

Let's start with `common_properties.xacro`. Here is the first thing to write inside this file:

```
<?xml version="1.0"?>
<robot xmlns:xacro="http://www.ros.org/wiki/xacro">
</robot>
```

All Xacro files must have this code, with a `<robot>` tag containing the `xmlns:xacro` attribute.

> **Note**
> Don't add the `name` attribute in the `<robot>` tag. This attribute will only be added once in the main Xacro file.

Then, inside the `<robot>` tag, you can copy and paste the two `<material>` tags we have previously written. You can then remove those tags from `my_robot.urdf.xacro`.

In `mobile_base.xacro`, you will also start the file with the `<?xml>` and `<robot>` tags, like we did in `common_properties.xacro`.

Then, you can copy and paste all the `<xacro:property>`, `<xacro:macro>`, `<link>`, and `<joint>` tags that are related to the mobile base, which are basically all the tags left.

In `my_robot.urdf.xacro`, we have nothing left, only the `<?xml>` tag and the `<robot>` tag that contains the `name` and `xmlns:xacro` attributes.

To include a Xacro file inside another file, you will write a `<xacro:include>` tag and provide the path to the file with the `filename` attribute. Here is the final content for `my_robot.urdf.xacro`:

```
<?xml version="1.0"?>
<robot name="my_robot" xmlns:xacro="http://www.ros.org/wiki/xacro">
    <xacro:include filename="common_properties.xacro" />
    <xacro:include filename="mobile_base.xacro" />
</robot>
```

Our URDF is now split into several files and uses Xacro properties and macros. With those modifications, we didn't change anything about the robot model, but we made the URDF more dynamic and scalable, as well as easier to read.

As a reminder, you can find the complete code—all URDF and Xacro files—for this chapter inside the `ch11` folder of the book's GitHub repository.

Summary

In this chapter, you discovered the full process of writing a URDF for a robot.

A URDF defines a robot model and contains two main things: links and joints. A link is a rigid part of a robot that does nothing on its own. A link can have a visual (simple shapes such as boxes, cylinders, spheres, or meshes exported from CAD software). You can see a robot as a collection of links put together. A joint defines the connection between two links. It specifies which link is the parent and which one is the child, as well as where the two links are connected, and how they move relative to each other.

You learned that what you write inside a joint will define a TF for the robot. In the end, with all the joints inside a URDF, you are creating a TF tree.

You also saw the complete process for adding a new link and joint on top of the previous ones. Make sure to follow this process every time. To help you develop and verify each step of the process, you learned that it's a good idea to use tools such as RViz to visualize the robot model, and `tf2_tools` to see the TF tree.

Then, you learned that you can also improve your URDF with Xacro. You can define some properties and macros, and even split a URDF into several files. This will be useful as your application scales and will make collaboration easier.

Creating a URDF for a robot is the first step, as mentioned in this chapter. This will allow you to generate the TFs, which are the backbone of any robot using ROS.

Now that you can create the URDF, let's see how to start packaging our application, and discover what to start so we can properly generate the TFs (without using the `urdf_tutorial` package). This will be the focus of the next chapter.

12
Publishing TFs and Packaging the URDF

So far, in *Part 3* of this book, you've had an introduction to TFs and have learned how to write a URDF, which will be used to generate the TFs for your robotics application. There are now two things we need to do to go further.

First, to be quicker, we have used the `urdf_tutorial` package to publish the TFs. This is great to get started and visualize the robot model, but we won't use this package in a real application. The question is this then: Using the URDF we have created, how do we generate the TFs for our application? What nodes do we need to start? We will first understand, through experimentation, what nodes and parameters we need to start to correctly generate the TFs for our application. From this, we will be able to create our own launch file.

Second, the URDF is now a series of three Xacro files placed inside the home directory. To start a proper ROS 2 application, we will create a package to organize and install the URDF, the launch file, and so on.

By the end of this chapter, you will be able to correctly package your URDF and publish the TFs for a ROS 2 application. This process is the same for any robot, and this package you create here will be the base for any further development, including the Gazebo simulation that we will cover later in this book.

As a starting point for this chapter, we will use the code inside the `ch11` folder from the book's GitHub repository (`https://github.com/PacktPublishing/ROS-2-from-Scratch`). You can find the final code in the `ch12` folder.

In this chapter, we will cover the following topics:

- Understanding how to publish TFs with our URDF
- Starting all nodes from the terminal
- Creating a package to install the URDF
- Writing a launch file to publish TFs and visualize the robot

Understanding how to publish TFs with our URDF

We will start this chapter by understanding what nodes and parameters we need to start in order to publish the TFs for our application. Then, with this knowledge, we will be able to start the required nodes, package the application, and write a launch file.

As we often did in this book, we will start with a discovery phase, through experimentation. I recommend that you run all the commands as you read the section.

The robot_state_publisher node

Basically, in this chapter, we want to replicate what was done with the `urdf_tutorial` package so we can publish the TFs by ourselves. Let's then start the `display.launch.py` launch file again, using the URDF from the previous chapter, and do some introspection:

```
$ ros2 launch urdf_tutorial display.launch.py model:=/home/<user>/
my_robot.urdf.xacro
```

In a second terminal, start `rqt_graph` to visualize the nodes that are currently running. Refresh the view if the screen is empty. You will see something like this:

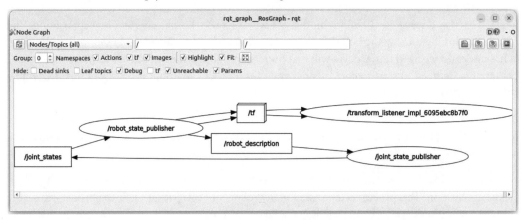

Figure 12.1 – Nodes running with urdf_tutorial

We see the `/tf` topic, which is the most important thing here. This is what's needed for any ROS application to work properly.

Now, what is publishing on the `/tf` topic? As you can see, there is a node called `/robot_state_publisher` (in the text, we will write `robot_state_publisher`, without the leading slash). Where does this node come from? `robot_state_publisher` is a core node already available for you to use. It is a part of the collection of packages that you installed with ROS 2. This node will publish the TFs for your robot. You will start it in any ROS 2 application where you need TFs. Most of the time, you won't have to publish any TF by yourself, as this will be handled by the `robot_state_publisher` node.

Now that we know we have to start this node, what inputs are required?

Inputs for the robot_state_publisher

There are two things you need to provide for the `robot_state_publisher` node to work correctly: URDF and joint states. Let's start with the first one.

URDF as a parameter

At this point, you might wonder: Where did the URDF go? We just saw some nodes and topics on `rqt_graph`, but we didn't see the use of the URDF we created.

Keep `display.launch.py` from `urdf_tutorial` running, and in another terminal, list all parameters for the `robot_state_publisher` node:

```
$ ros2 param list /robot_state_publisher
```

You will see quite a few, but the one that we care about here is named `robot_description`. Then, you can read the value from this parameter:

```
$ ros2 param get /robot_state_publisher robot_description
```

With this, you will see the entire URDF in the terminal (or more precisely, the generated URDF from the Xacro file you have provided).

So, when you start the `robot_state_publisher` node, you will need to give the URDF inside a parameter named `robot_description`.

> **Note**
> In `rqt_graph`, you can see that the `robot_state_publisher` is publishing on the `/robot_description` topic. The message you get from this topic also contains the URDF content. This can be useful to retrieve the URDF from any other node, using a subscriber.

That's it for the first input; let's see the second one.

Joint states topic

In order to publish the TFs, the `robot_state_publisher` node will need the URDF, but also the current state for each joint.

You can see the `/joint_states` topic in `rqt_graph` in *Figure 12.1*. This topic contains what you would read from encoders or control feedback in a real robot. For example, if you have some wheels, you will get to know the speed and/or position of those wheels. You will feed this into the `/joint_states` topic. If you have a robotic arm, you usually have encoders on each axis reading the current position for the axis.

When we simulate the robot with Gazebo, we will use a plugin that automatically publishes the joint states. In fact, either in simulation mode or for a real robot, you will usually have nodes doing this for you (to go further on this, check out ros2_control after reading this book—you can find extra resources about that in the last chapter of this book). So, all you need to know is that this `/joint_states` topic is important, as it's required by `robot_state_publisher`. Publishing on this topic is done by existing ROS 2 plugins.

For now, as we don't have any real robot or Gazebo simulation, we will use the `joint_state_publisher` node (with the **Joint State Publisher** window), which will publish whatever values we select on the cursors. For example, if you select `1.0` radian for `base_right_wheel_joint`, then `1.0` will be published on the `/joint_states` topic for that joint and will be received and used by `robot_state_publisher`.

Here, it's important to clarify the difference between joint states and TFs. A joint state is simply the *current state* of a joint. For example, for a wheel: what is the current velocity? For an axis in a robotic arm: what is the current angular position of the motor? The state is one data point, at one specific time, for one joint in your robot. It doesn't specify anything about the relationship between joints, nor where they are located relative to each other—this is what a TF is.

Recap – how to publish TFs

Let's do a quick recap of what is required to publish TFs. This is what `urdf_tutorial` was doing for us, and what we will be doing by ourselves from now on.

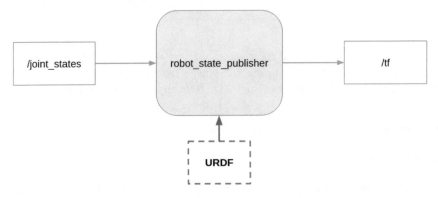

Figure 12.2 – The required node and inputs to publish TFs

Here is what we need to do:

1. Start the `robot_state_publisher` node. This node is already installed. We give the URDF as the `robot_description` parameter.
2. Publish the current state for all joints on the `/joint_states` topic. This is usually done automatically for you, either with data from encoders, a simulator such as Gazebo, or *fake* data from the `joint_state_publisher` node.

That's all you have to do to correctly publish the TFs for your application. Those TFs will then be used by other nodes, plugins, and stacks, for example, the Navigation 2 or MoveIt 2 stacks (we won't cover those in this book, but they would be a good topic to study afterward—more resources will be given in the last chapter of this book).

Starting all nodes from the terminal

Before we package the application and write a launch file, let's start all the nodes we need in the terminal. Doing this is a best practice so that you can make sure your application is working properly. Then, creating the package and the launch file will be easier, as you already know all the elements you need to include.

We will use the result from the previous section and start the `robot_state_publisher` node, as well as the `joint_state_publisher` node. In addition to that, we will start RViz (this is optional and only used to visualize the robot model).

Publishing the TFs from the terminal

Let's publish the TFs. For that, we will open two terminals.

In the first one, start the `robot_state_publisher` node. The package and executable names for this node are identical. To provide the `robot_description` parameter, you will have to use this syntax: `"$(xacro <path_to_urdf>)"`.

In Terminal 1, run the following command:

```
$ ros2 run robot_state_publisher robot_state_publisher --ros-args -p
robot_description:="$(xacro /home/<user>/my_robot.urdf.xacro)"
[robot_state_publisher]: got segment base_footprint
[robot_state_publisher]: got segment base_link
[robot_state_publisher]: got segment caster_wheel_link
[robot_state_publisher]: got segment left_wheel_link
[robot_state_publisher]: got segment right_wheel_link
```

If you see this, everything is working fine. `robot_state_publisher` has been started with the URDF, and it's ready to publish on the `/tf` topic. Now, we need to add a publisher on the `/joint_states` topic. We will use the `joint_state_publisher_gui` executable from the same package (note the extra `_gui` suffix, which means **graphical user interface**).

> **Note**
>
> As a reminder, the executable name and node name are two different things. The executable name is the one you define in `setup.py` (for Python) or `CMakeLists.txt` (for C++). The node name is defined in the code and can be different. Here, we start the `joint_state_publisher_gui` executable, but the node name is `joint_state_publisher`.

In Terminal 2, run the following command:

```
$ ros2 run joint_state_publisher_gui joint_state_publisher_gui
```

This will open the **Joint State Publisher** window that we previously used when experimenting with TFs and URDF. The values you see on the cursors will be published on the `/joint_states` topic and will be received by `robot_state_publisher`.

That's basically all we need. This will become the backbone of your ROS 2 application—of course, we will need to package this nicely and start it from a launch file.

If you run `rqt_graph`, you will see the same nodes and topics as in *Figure 12.1*. You can also print the TF tree (`ros2 run tf2_tools view_frames`) and listen to the `/tf` topic in the terminal (`ros2 topic echo /tf`).

Visualizing the robot model in RViz

On top of what we did, we can visualize the robot in RViz. This is optional, and it's something you will do mostly when you develop your application. After everything is working correctly and you switch to production mode, you won't need to start RViz.

Starting and configuring RViz

Let's start RViz and see how to visualize the robot model as well as the TFs.

Keep the `robot_state_publisher` and `joint_state_publisher` nodes running. Then, in Terminal 3, run the following command:

```
$ ros2 run rviz2 rviz2
```

This will open RViz, but, as you can see, there is no robot model, and we have some errors on the left menu:

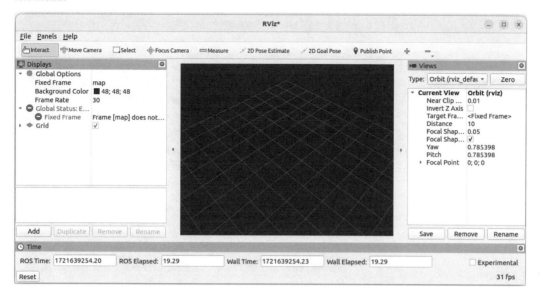

Figure 12.3 – RViz with no robot model and some errors

We need to do a bit of configuration to correctly visualize the robot model and the TFs. Then, we will be able to save this configuration and reuse it the next time we start RViz.

Follow these steps to configure RViz:

1. In the left menu, **Global Options | Fixed Frame**, change from `map` to `base_footprint`. After that, `Global Status: Error` should change to `Global Status: OK`.

2. Click on the **Add** button on the left, scroll down, and double-click on **RobotModel**. You will have a new menu on the left side of RViz.

3. Open this new **RobotModel** menu, find **Description Topic**, and click on the empty space on the right side of the menu (this one is a bit tricky to find). You should see a drop-down menu; here, select `/robot_description`. After this, the robot model should appear on the screen.

4. Click on the **Add** button again, scroll down, and double-click on **TF**. This will open a new menu, and you will see the TFs appear on the screen.

5. If you want to see through the model, like we did before, open **RobotModel**, and reduce the **Alpha** (transparency) value from `1` to `0.8`, for example.

6. You can remove the extra menus on the right (**Views**) and at the bottom (**Time**) to get more space for the robot.

With all those settings, you should see the robot model and TFs the same way we did when we previously visualized the URDF with the `urdf_tutorial` package:

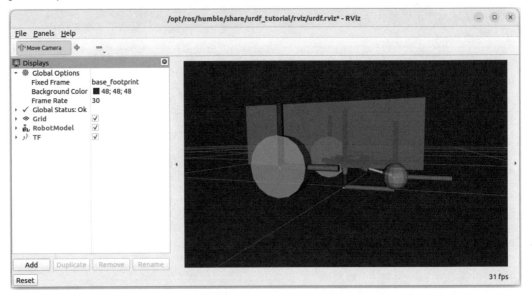

Figure 12.4 – RViz with the robot model and TFs

Saving the RViz configuration

You will need to repeat those steps every time you start RViz. To avoid doing this, we will save the configuration.

Click on **File | Save Config As**. Let's name the file `urdf_config.rviz` (for these files, use the `.rviz` extension), and place it inside your home directory for now.

Make sure you can see the file, using a file manager or the terminal. If you didn't save the file correctly, you will need to manually do the full configuration again. Once the file is saved, you can stop RViz (*Ctrl* + *C* in the terminal).

Then, when you start RViz again, you can add an extra `-d` argument with the path to the configuration file:

```
$ ros2 run rviz2 rviz2 -d /home/<user>/urdf_config.rviz
```

This will start RViz exactly like you saved it: same menus, same view, same zoom, and so on. We will reuse this configuration file throughout this chapter.

> **Note**
> If you wish to change the configuration, all you have to do is modify whichever settings you want in RViz, save a new configuration file, and use this one instead.

We now have everything we need: the URDF files and the RViz config file, and we know what nodes and parameters we have to start, and how to start them.

Let's now organize everything properly into a ROS 2 package. We will first create the package, and then add a launch file to start all the nodes at once.

Creating a package to install the URDF

All the files we have created are now in our home directory. It's time to create a ROS 2 package and move all the files into the right place so that they can be installed and used in our ROS 2 workspace.

We will start by creating a package dedicated to the robot model. Then, we will install all the files required for this application. This will allow us to use the URDF and RViz files when we write a launch file, to start all the nodes we have seen previously.

Let's create a new package, but before we do that, it could be a good idea to create a new ROS 2 workspace.

Adding a new workspace

As of now, our `ros2_ws` workspace contains all the code used in *Part 2* of this book, including various examples to illustrate the core concepts, a robot controller for the Turtlesim, some custom interfaces, and launch files. We don't need any of those for the *Part 3* project; so, instead of continuing to add more things to this workspace, we will create a new one. As a general rule, if you have two different applications, you will have two different workspaces.

Let's then create a new workspace named `my_robot_ws`. A good practice is to name the workspace as per your application or robot name. This will help you avoid confusion in the long term.

Create a new workspace in your home directory:

```
$ mkdir ~/my_robot_ws
```

Then, inside this workspace, create a `src` directory:

```
$ mkdir ~/my_robot_ws/src
```

Now, and this is super important, you can have as many ROS 2 workspaces as you want, but you should not work with two workspaces for two different applications at the same time.

You are currently sourcing the `ros2_ws` workspace every time you open a terminal. If you remember, you added an extra line in `.bashrc` to do that:

```
source ~/ros2_ws/install/setup.bash
```

Open the `.bashrc` file again and comment that line (add # in front of the line). Now, close all terminals, open a new one, and let's build and source our new workspace:

```
$ cd ~/my_robot_ws/
$ colcon build
```

Open the `.bashrc` again and add a line to source this new workspace. The end of `.bashrc` will look like this:

```
source /opt/ros/jazzy/setup.bash
#source ~/ros2_ws/install/setup.bash
source ~/my_robot_ws/install/setup.bash
```

As a reminder, we first source the global ROS 2 installation. Then, we source our workspace. Commenting the workspaces you don't use is very practical. This way, if you want to switch between two workspaces, you just have to uncomment/comment the two lines, close all terminals, and open a new terminal to source the workspace you want to use.

Creating a _description package

Now that we have this new empty workspace correctly configured and sourced, let's add a new package inside.

To name the package, we use the robot's name, followed by `_description`. This is a common convention used by lots of ROS developers. By using this naming format, you make it clear that this package will contain the URDF files for your robot. This will make collaboration easier.

To create this new package, we use the `ament_cmake` build type, just as if we were creating a C++ package, and we don't specify any dependencies. Let's create the package:

```
$ cd ~/my_robot_ws/src/
$ ros2 pkg create my_robot_description --build-type ament_cmake
```

For now, this package is a standard C++ ROS 2 package. However, we won't write any nodes inside. We just need the package to install our robot model. Thus, you can remove the `src` and `include` directories:

```
$ cd my_robot_description/
$ rm -r include/ src/
```

At this point, your package will only contain two files: `package.xml` and `CMakeLists.txt`.

Open the workspace with an IDE. If you're using VS code, run the following:

```
$ cd ~/my_robot_ws/src/
$ code .
```

To clean things a bit more, we will simplify the `CMakeLists.txt` file. Remove the comments after the `find_package(ament_cmake REQUIRED)` line, and, as we don't need that now, remove the `if(BUILD_TESTING)` block. Make sure you keep the `ament_package()` instruction, which should be the last line of the file.

Installing the URDF and other files

Now that we have our `robot_description` package, let's install all the files we need. To install a file, we will follow this process:

1. Create a folder to host the files.
2. Add an instruction in `CMakeLists.txt` to install the folder.

Here, we will see how to install the URDF files, custom meshes, and the RViz configuration.

Installing Xacro and URDF files

To install our Xacro and URDF files, go inside the `my_robot_description` package and create a new folder named `urdf`:

```
$ cd ~/my_robot_ws/src/my_robot_description/
$ mkdir urdf
```

You can now move all three Xacro files inside this `urdf` folder: `common_properties.xacro`, `mobile_base.xacro`, and `my_robot.urdf.xacro`.

Everything should work fine, but to make the `include` paths more robust, let's modify the `<xacro:include>` tags inside `my_robot.urdf.xacro`:

```
<xacro:include filename="$(find my_robot_description)/urdf/common_properties.xacro" />
<xacro:include filename="$(find my_robot_description)/urdf/mobile_base.xacro" />
```

Instead of providing only the relative path (which should still work in this case), we provide the absolute path to where the files will be installed, using the `find` keyword before the package name.

Now, open `CMakeLists.txt`, and add those instructions to install the `urdf` folder:

```
find_package(ament_cmake REQUIRED)
install(
  DIRECTORY urdf
```

```
    DESTINATION share/${PROJECT_NAME}/
)
ament_package()
```

This will install the `urdf` folder inside a `share` directory when you build the package. It will allow any package from the workspace to find the URDF for your robot. This is actually very similar to what we've done in *Chapter 9* when installing launch files and param files.

Let's now continue with custom meshes.

Installing custom meshes

If you are not using any custom mesh for now (which is probably the case if you are learning ROS 2 from scratch with this book), you can skip this small section and go to the RViz configuration directly.

I briefly mentioned custom meshes in *Chapter 11*, when we were creating links for the URDF. If you happen to include custom meshes with the `<mesh>` tag, inside the `<visual>` tag of a link, you will have to install the corresponding mesh files (with the `.stl` or `.dae` extension).

In this case, in `my_robot_description`, you would create a new folder named `meshes`:

```
$ cd ~/my_robot_ws/src/my_robot_description/
$ mkdir meshes
```

In this folder, you would add all the `.stl` and `.dae` files you want to use in your URDF. Then, let's say you have added a file named `base.stl`. In the URDF, you will use this syntax to include it:

```
<mesh filename="file://$(find my_robot_description)/meshes/base.stl"
/>
```

To install the `meshes` folder, you also have to add an instruction in `CMakeLists.txt`. As we already added the `install()` block previously, you just need to add the name of the new folder you want to install:

```
install(
  DIRECTORY urdf meshes
  DESTINATION share/${PROJECT_NAME}/
)
```

> **Note**
> When adding new folder names after `DIRECTORY`, you can either separate them with a space or put them on a new line; it won't make a difference.

That's it for custom meshes. Even though we don't use them now, I wanted to include this, so that you have all the required knowledge to create a complete URDF with custom shapes.

Installing the RViz configuration

Let's finish this section by installing the RViz configuration we previously saved. This way, when we start RViz later on from a launch file, we can use this configuration directly from the package.

Create an `rviz` folder inside the package:

```
$ cd ~/my_robot_ws/src/my_robot_description/
$ mkdir rviz
```

Move the `urdf_config.rviz` file into this new `rviz` folder.

Now, to install the folder, add its name to the `install()` instruction, inside `CMakeLists.txt`:

```
install(
  DIRECTORY urdf meshes rviz
  DESTINATION share/${PROJECT_NAME}/
)
```

We now have all the files we need for this package. Before we install them (with `colcon build`), let's add a launch file so we can start all the required nodes to publish the TFs as well as visualize the robot model in RViz.

Writing a launch file to publish TFs and visualize the robot

The `my_robot_description` package is finished, but we will add a launch file so we can start all the nodes and parameters that we discovered at the beginning of this chapter. This way, we can publish TFs and visualize the robot in RViz. This will also be a good practice exercise on launch files, and we will reuse part of the code in the next chapter when we build the Gazebo simulation for the robot.

> **Note**
> Usually, we add all launch files inside a `_bringup` package (dedicated package used for launch and configuration files). Here, we make an exception, because this launch file will be used only for visualization and development. Any other launch file that we write for this application will be placed inside the `my_robot_bringup` package (which we will create in the next chapter).

We will write the launch file first with XML, and then with Python. This will be another example of how XML launch files can be easier to write than Python ones.

The XML launch file

Before writing any launch file, we first need to create a `launch` folder, where we will put all our launch files for the `my_robot_description` package.

Creating and installing a launch folder

To create and install a `launch` folder, we will follow the same process as before. First, create the folder:

```
$ cd ~/my_robot_ws/src/my_robot_description/
$ mkdir launch
```

Then, to specify the instructions to install this folder, it's as easy as adding the folder name inside `CMakeLists.txt`:

```
install(
  DIRECTORY urdf meshes rviz launch
  DESTINATION share/${PROJECT_NAME}/
)
```

As the `install()` instruction was already configured, if we want to install a new folder, we just add its name after the other ones. For now, we are installing four folders in this package.

Once this is done, create a new launch file inside the folder. We will name it `display.launch.xml`:

```
$ cd launch/
$ touch display.launch.xml
```

Now, let's write the content for this launch file.

Writing the launch file

In this launch file, we will simply start the nodes and parameters that we discovered and listed at the beginning of this chapter:

- `robot_state_publisher` with the URDF as the `robot_description` parameter. This node will publish on the `/tf` topic.
- `joint_state_publisher` to publish on the `/joint_states` topic.
- `rviz2`, as we also want to visualize the robot model.

We already know how to start those nodes from the terminal; now, all we have to do is add them inside one launch file.

In the launch file, first open and close a `<launch>` tag:

```
<launch>
</launch>
```

Now, make sure to write all the following lines inside this `<launch>` tag. You can also add an indentation (four spaces) for those lines.

As we will need to find the path of the URDF file and the RViz configuration file, we add two variables at the beginning of the launch file, to make the file cleaner and more readable. Also, if you have to modify those values later, you know that you just have to modify the variables at the top of the file.

We haven't seen how to add (constant) variables yet in a launch file, but it's not too complicated. You will use the `<let>` tag with two arguments: `name` and `value`. Let's add the two variables we need:

```
<let name="urdf_path" value="$(find-pkg-share my_robot_description)/
urdf/my_robot.urdf.xacro" />
<let name="rviz_config_path" value="$(find-pkg-share my_robot_
description)/rviz/urdf_config.rviz" />
```

Then, to use a variable, you can write `$(var name)`.

> **Note**
>
> Although the XML syntax looks similar between launch files and Xacro files, make sure not to mix things up: *To find a package*, you use `find-pkg-share` in a launch file and `find` with Xacro. *To use a variable*, you use `$(var name)` in a launch file and `${name}` with Xacro.

Now, let's start all the nodes we need, one by one. Here is the code for the `robot_state_publisher` node:

```
<node pkg="robot_state_publisher" exec="robot_state_publisher">
    <param name="robot_description"
            value="$(command 'xacro $(var urdf_path)')" />
</node>
```

Nothing special here; we use the same values as in the command we previously ran in the terminal. To specify a command to run in an XML launch file, you can use `$(command '...')`.

Next, we can start the `joint_state_publisher` node. Here, we use the executable with the `_gui` suffix to get a graphical window with cursors to move the joints:

```
<node pkg="joint_state_publisher_gui"
      exec="joint_state_publisher_gui" />
```

This node is easy to write, as we don't need to provide anything else except the package and executable. Let's finish with the RViz node:

```
<node pkg="rviz2" exec="rviz2" args="-d $(var rviz_config_path)" />
```

In this node, we provide the saved RViz configuration file, using the `-d` option.

Starting the launch file

The launch file is now complete. We can build the workspace to install all the files and folders that we've added:

```
$ cd ~/my_robot_ws/
$ colcon build --packages-select my_robot_description
```

Then, don't forget to source the workspace (`source install/setup.bash`), and you can start your new launch file:

```
$ ros2 launch my_robot_description display.launch.xml
```

After running this command, you should see your robot model in RViz with the TFs. If you list nodes and topics on the terminal, you should be able to find everything that we have started. With `rqt_graph`, you should also get the same result as when we started all three nodes from the terminal.

The `my_robot_description` package is now complete. We have installed all the files that we need: URDF, custom meshes, RViz configuration, and a launch file to publish TFs and visualize the robot model. Before we wrap things up, let's talk briefly about the Python version of the launch file.

The Python launch file

We will finish this chapter with the Python launch file to do the same thing: start the `robot_state_publisher`, `joint_state_publisher`, and RViz. The main reason I do this here is to provide another example to emphasize the difference between Python and XML launch files. You can see this as an extension of what we discussed during *Chapter 9, XML versus Python for launch files*.

This is also the last time I use Python for launch files, as in the next chapter on Gazebo, we will focus only on XML (as a reminder: If you ever need to use an existing Python launch file, you can include it inside an XML launch file, so there is no problem here).

Let's now create the Python launch file. To do that, create a new file inside the `launch` folder of the `my_robot_description` package. You can name this file `display.launch.py`—the same name as for the XML launch file, but with a Python extension.

There is no need to add any extra configuration in the package, as `CMakeLists.txt` already contains the instructions to install the `launch` folder.

Let's start to analyze this Python file (you can find the full code for this file in the book's GitHub repository) with the import lines:

```
from launch import LaunchDescription
from launch_ros.parameter_descriptions import ParameterValue
from launch_ros.actions import Node
```

```
from launch.substitutions import Command
import os
from ament_index_python.packages import get_package_share_path
```

As you can see, that's a lot of things to import, and it could be easy to make mistakes here. After the imports, we start the `generate_launch_description` function:

```
def generate_launch_description():
```

All the code we have from now on will be inside this function (with an indentation). We create the two variables for the URDF and RViz path:

```
urdf_path = os.path.join(
    get_package_share_path('my_robot_description'),
    'urdf',
    'my_robot.urdf.xacro'
)

rviz_config_path = os.path.join(
    get_package_share_path('my_robot_description'),
    'rviz',
    'urdf_config.rviz'
)
```

To make things a bit more readable, we also create a variable for the `robot_description` parameter:

```
robot_description = ParameterValue(Command(['xacro ', urdf_path]),
value_type=str)
```

Now, we start the three nodes:

```
robot_state_publisher_node = Node(
    package="robot_state_publisher",
    executable="robot_state_publisher",
    parameters=[{'robot_description': robot_description}]
)

joint_state_publisher_gui_node = Node(
    package="joint_state_publisher_gui",
    executable="joint_state_publisher_gui"
)

rviz2_node = Node(
    package="rviz2",
    executable="rviz2",
    arguments=['-d', rviz_config_path]
)
```

Finally, we need to return a `LaunchDescription` object with all the nodes we want to start:

```
return LaunchDescription([
    robot_state_publisher_node,
    joint_state_publisher_gui_node,
    rviz2_node
])
```

As you can see, the code is much longer: 38 lines for Python versus 15 lines for XML. We could optimize the spaces, but even with that, there would still be a factor of two. Also, I personally find the Python syntax more complex and not really intuitive. To be honest with you, the only way I manage to write Python launch files is just by scrapping bits of code from existing projects I find on GitHub and seeing if it works.

I'm not going to provide much more details than that, as this code was here mainly so we can have another example of what we discussed back in *Chapter 9*. We will continue with XML launch files going forward.

Summary

In this chapter, you have published the TFs for your robot and properly packaged your application.

You first discovered that the most important node to start is `robot_state_publisher`, with two inputs: the URDF in the `robot_description` parameter, and the current states for each joint on the `/joint_states` topic.

From this, you started all the nodes and parameters from the terminal so that you could reproduce the output we previously had with the `urdf_tutorial` package.

Then, you created your own package to correctly organize your application. Here is the final architecture for this package, with all the files and folders:

```
~/my_robot_ws/src/my_robot_description
├── CMakeLists.txt
├── package.xml
├── launch
│   ├── display.launch.py
│   └── display.launch.xml
├── meshes
├── rviz
│   └── urdf_config.rviz
└── urdf
    ├── common_properties.xacro
    ├── mobile_base.xacro
    └── my_robot.urdf.xacro
```

This organization is very standard. If you look at the code for almost any robot powered by ROS, you will find this `_description` package using more or less the same architecture. Thus, creating this package was not only helpful for the process we follow in this book but also for you to easily start working on any other project made by ROS developers.

We will now use this package as the base for the next chapter, where we will learn how to create a simulation for the robot using Gazebo.

13
Simulating a Robot in Gazebo

In the previous chapters, you wrote a URDF to describe a robot, published the TFs for that robot, and properly organized all files into the `my_robot_description` package.

You are now going to simulate the robot in Gazebo. This will be the end of the *Part 3* project. The goal here is to finish the book with a working simulation. After that, I will conclude by giving you some hints on what to do to further with ROS.

We will start the chapter by understanding what Gazebo is, how it is integrated with ROS 2, and how to work with it. This will allow us to adapt the robot URDF for Gazebo, spawn it in the simulator, and control it with a plugin. We will also properly package the application so that we can start everything from one single launch file.

By the end of this chapter, you will be able to simulate a robot in Gazebo and interact with it using ROS 2. After you've done the process once, it will be much easier for the next robot you want to simulate.

The level of this chapter is more advanced than what we did before. We will reach a point where the documentation lacks a lot. Finding useful information usually means doing a lot of research on Google, as well as finding GitHub code that works and that you can use as an inspiration.

To complete this chapter, you will also need to utilize what you have previously learned from this book—for example, creating and organizing packages, working with topics and parameters, and writing launch files. Don't hesitate to refer to previous chapters if you have any doubts.

We will use the code inside the `ch12` folder (on the GitHub repository at https://github.com/PacktPublishing/ROS-2-from-Scratch) as a starting point. You can find the final code in the `ch13` folder.

In this chapter, we will cover the following topics:

- How Gazebo works
- Adapting the URDF for Gazebo
- Spawning the robot in Gazebo
- Controlling the robot in Gazebo

Technical requirements

If you installed ROS 2 in a **VM** (as explained with VirtualBox at the beginning of the book), it probably worked well for all chapters in *Part 1* and *Part 2*, and possibly for the previous chapters in *Part 3*, even when running RViz.

However, with Gazebo, chances are that the VM won't be enough. VirtualBox doesn't work well with 3D simulation tools. From now on, I would strongly recommend that you have Ubuntu installed with a dual boot. I know that some people had more success with VMware Workstation (using the free version for personal use) or WSL 2 on Windows. If you find another combination that works for you, fine, but I would still recommend a dual boot, which will possibly be less buggy and bring you a better experience overall.

So, if you are currently running a VM, take the time to set up a dual boot and install Ubuntu 24.04. Then, follow the instructions from *Chapter 2* again to install ROS Jazzy. This will take you a bit of time in the short term, but it will surely be more efficient in the long term.

How Gazebo works

Before we work on our application, it's important to understand what Gazebo is and how it works.

Gazebo is a 3D simulation engine. It contains a physics engine (with gravity, friction, and other physical constraints) with which you can simulate a robot, just like if it were in the real world.

That's one of the strengths of Gazebo. You can develop your application using mostly the Gazebo simulation and then work with the real robot. This brings a lot of benefits. For example, you can work on robots that don't exist yet, test extreme use cases without damaging the real robot, create custom environments you can't access on a daily basis, work remotely, and so on.

In this section, we will start Gazebo and explore a few functionalities. We will also see how to connect Gazebo with ROS 2 and understand the steps we need to take to adapt our robot for Gazebo. Before getting started with this, a common question that lots of people have at this point is, what is the difference between Gazebo and RViz?

Clarifying – Gazebo versus RViz

We are starting with this because I believe this is one of the first confusions you can have. We have already used RViz previously, and we could visualize the robot in it, as well as lots of other information. So, why do we need Gazebo?

To understand this, let's first go back to what RViz is. RViz is a 3D visualization tool. With RViz, you can visualize the URDF, the TFs, and the data you get from ROS topics. This is a great tool to use when developing, as you can check that what you are doing is correct.

Now, RViz **is not** a simulation tool. You don't simulate anything; you **only visualize** what already exists. So, the robot and all the data you see in RViz is only the representation of what's happening externally (of RViz). Instead of seeing and interacting with all the data in the terminal, you can do that with a graphical interface and see the data in 3D. That's (very simply put) what RViz brings to you.

For example, with TFs, RViz will subscribe to the /tf topic and display the TFs on the screen. However, RViz doesn't control the TFs; it just shows them.

Conversely, Gazebo is a simulation tool. It will simulate gravity and the real physical properties of the robot. It also has some control plugins so that you can simulate the hardware control, and even publish the joint states and TFs for your robot.

So, what do we use here—Gazebo or RViz? Ultimately, it's not a competition; both are complementary.

Gazebo is available if you don't have a real robot, if you don't want to use it, or if you want to test a robotics system in a different environment, for example. With Gazebo, you can replicate the behavior of your robot, and make it very close to what it would do in real life.

On top of Gazebo, you can use RViz to visualize the TFs and other important data from your application. RViz is still a very useful tool to have during the development phase. So, the dilemma is not *Gazebo versus RViz*, but instead, *real robot versus Gazebo*. In the end, you can have either:

- Real robot and RViz
- Gazebo simulation and RViz

In this chapter, we will use the second combination. We will spawn and control our robot in Gazebo. Then, we will also visualize it in RViz. What we see in RViz will be a reflection of what's happening in Gazebo. If, later on, we switched to a real robot, we would ditch Gazebo but still use RViz to see whether everything still works well.

Now that we have made this clarification, we can get started with Gazebo.

Starting Gazebo

Let's run Gazebo and get used to the interface.

First, you need to install Gazebo. We installed ROS 2 previously with `ros-<distro>-desktop`, which already contains a lot of packages. To get everything we need for Gazebo, install this additional package:

```
$ sudo apt install ros-<distro>-ros-gz
```

Make sure to replace `<distro>` with your current ROS 2 distribution, and after running this command, source your environment.

> **Note**
>
> Since we are using Ubuntu 24.04, this will install **Gazebo Harmonic**. For other Ubuntu versions, you can find the recommended Gazebo versions here: https://gazebosim.org/docs/latest/getstarted/.

It's important to note that Gazebo is actually independent from ROS 2. You can run Gazebo on its own, without ROS 2. In fact, you could even install Gazebo without any ROS 2 packages.

Gazebo has been designed as an independent robotics simulator that you can then use with ROS 2, as well as other robotics frameworks. Here, I will only focus on ROS 2. The reason I mention this is just to make you understand that Gazebo and ROS are separate projects.

We will start by running Gazebo on its own. Then, we will see how to connect Gazebo and ROS, which will be useful for planning the steps to adapt our project and simulate our ROS robot on Gazebo.

So, to start Gazebo (without ROS), run this command in a terminal:

```
$ gz sim
```

You will be taken to a Gazebo quick-start menu. There, you can click on **Empty** to load an empty world. You also have other existing worlds (click on the different images, or use the search bar) that you can explore later on your own. Warning—some of these worlds may contain bugs and not work properly.

To stop the simulation, press *Ctrl + C* in the terminal where you started Gazebo.

When running the Gazebo command, you can also directly specify the world you want to launch. World description files in Gazebo use the SDF format (a file with a `.sdf` extension), which is quite similar to URDF. Let's start Gazebo with the empty world:

```
$ gz sim empty.sdf
```

Now, take the time to move around the space using mouse controls. You can also use a laptop touchpad, but if possible, I really recommend that you have a mouse, which will make things easier going forward.

At the bottom left of the screen, you will see a play button. Click on it to start the simulation.

After starting the simulation, you will see a percentage at the bottom-right part of the screen, representing the real-time factor. Basically, this will tell you immediately whether your computer is powerful enough for Gazebo. Personally, I have a real-time factor of about 98%, which means that the simulated time in Gazebo keeps up with real time at 98% speed (i.e., after 100 real seconds, 98 seconds have passed in Gazebo). If your percentage is too low, it may be a sign that you need better performance. If you're not sure, continue with the chapter, and you will quickly see whether the robot simulation works properly or not.

> **Note**
>
> Gazebo can be quite buggy, so don't be surprised if it crashes at some point—even on a powerful computer. If you can't close Gazebo properly (with *Ctrl + C* in the terminal), you may have some trouble when starting it again. In this case, you can try to stop all Gazebo processes that might still be running in the background. To do that, run `ps aux | grep gz` to find all related processes. You will find a `pid` with four numbers for each `gz` process (if any). To stop a process, run `kill -9 <pid>`. If nothing works, the best thing to do is to restart your computer.

Let's return to the simulation we started—at the top of the screen, you can click on the different shapes and add them to the empty space. Take some time to experiment with this. Add a box into the space. Find the translation mode and rotation mode. Move the box around (especially on the *z* axis) and see what happens. If you lift the box up, then you should see the box falling down on the floor. This is because of the gravity and physical properties of the box.

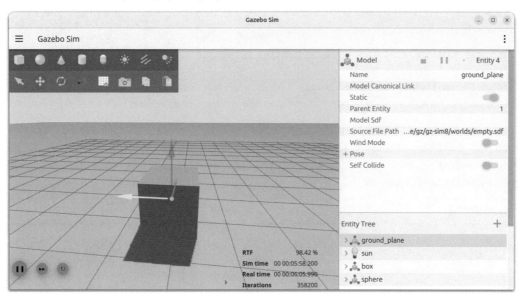

Figure 13.1: The Gazebo simulator with a box in an empty world

What we can do next is explore the communications used by Gazebo. Gazebo also uses topics and services, but those are not the same as ROS 2 topics and services.

For example, you can list all Gazebo topics with this command:

```
$ gz topic -l
```

You will see a lot of topics, but if you try to run `ros2 topic list` in another terminal, none of those topics will appear. Also, when running `ros2 node list`, you won't see any nodes.

With this, you can see that Gazebo is completely independent of ROS 2, and they do not interact with each other.

How Gazebo works with ROS 2

We will now explore how Gazebo and ROS 2 can be connected.

First of all, you can start Gazebo using a ROS 2 launch file from the `ros_gz_sim` package. This will be more practical for us because when we write our own launch file, we can include this one:

```
$ ros2 launch ros_gz_sim gz_sim.launch.py
```

This will start Gazebo the same way we did with the `gz sim` command. You can also specify the world to launch with the `gz_args` argument:

```
$ ros2 launch ros_gz_sim gz_sim.launch.py gz_args:=empty.sdf
```

However, even if we started Gazebo from a ROS 2 launch file, Gazebo is still independent. Try to list all the nodes and topics in a terminal; you will see the same result as before.

To connect Gazebo and ROS 2 topics (or services), you need to create a bridge between them. The `ros_gz_bridge` package does that for us, so we will use this package. We will only need to provide some configuration to specify which topics we want to bridge; how to write this configuration will be covered later in the chapter.

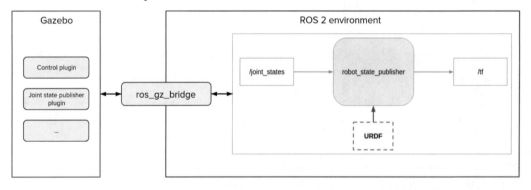

Figure 13.2: Connecting Gazebo and ROS 2 with ros_gz_bridge

In *Figure 13.2*, you can see the following:

- On the right, our current ROS 2 application with the `robot_state_publisher` node, publishing on the `/tf` topic.

- On the left, `Gazebo`. Inside `Gazebo`, we will add plugins (also called systems) to simulate the hardware behavior of the robot. For example, you could have one plugin to control the two wheels, and one plugin to publish the joint states for the wheels. This is what we will implement in this chapter.

- Then, to make everything work together, we will use the `ros_gz_bridge` package. With the joint state example, the Gazebo Joint state publisher plugin will publish the joint states with a Gazebo topic. Using `ros_gz_bridge`, we will match this topic with the ROS 2 `/joint_states` topic.

The important thing to understand here is that Gazebo and ROS 2 exist in two different environments, but you can make them work together. Your ROS 2 application will be the same, whether you work on a Gazebo simulation or a real robot. If you work on a real robot, then you would directly control the wheels and get the joint state data from encoders. With Gazebo, you use plugins to simulate the hardware.

Now, here are the steps we will take in the following sections to create the Gazebo simulation for our robot:

1. Adapt the URDF for Gazebo. For a robot to work on Gazebo, we first need to provide inertial and collision properties in the URDF.
2. Once the URDF is correct, we will start Gazebo and spawn the URDF in it. At this point, we will also create a package with a launch file.
3. We will then add some plugins (systems) to control the robot, using the `ros_gz_bridge` package to make those plugins communicate with our ROS 2 application.

You can follow this process for basically any ROS 2 robot you want to simulate in Gazebo. If some things are still not clear, continue with the chapter, work on the implementation, and come back to this section at the end. Everything will make more sense.

Let's start with the first step.

Adapting the URDF for Gazebo

We could try to spawn our robot directly into Gazebo, but it won't work, as the URDF is missing two key elements—inertial and collision properties. Gazebo needs those to correctly simulate a robot.

Thus, before we do anything with Gazebo, we need to come back to our URDF and add those properties. For each link of the robot that represents a physical part, we will add an `<inertial>` tag and a `<collision>` tag. In this section, you will learn how to properly configure those.

We will modify the URDF we created in the `my_robot_description` package. As a quick recap, this package contains the URDF, a launch file to display the robot model in RViz, and an RViz configuration. The most important thing in this package is the URDF. The launch file will help us validate that the values we set in the URDF are correct.

Let's start with the inertial properties and then add the collision ones.

Inertial tags

A URDF without inertial properties won't load in Gazebo. Thus, this is the first thing you need to add.

An `<inertial>` tag will contain a few elements, including a 3x3 matrix representing an inertia tensor. In this book, we won't dive into the theoretical inertia details; you can look that up on your own if you want, as there is pretty good documentation on the internet. Instead, we will focus on finding the correct formulas and applying them, allowing us to spawn the robot and quickly move on to the next steps.

So, currently, our URDF is split into three files. We will add some code to those files:

- `common_properties.xacro`: Here, we will add some macros to specify the inertial properties for a box, a cylinder, and a sphere. This way, we only need to write the inertial formulas once for those shapes, and you can reuse them in any of your projects.
- `mobile_robot.xacro`: Inside each link representing a physical part, we will use the corresponding inertial macro we defined previously.
- `my_robot.urdf.xacro`: Nothing changes here; we still import the two previous files.

Now, how do we write the inertial macros for the shapes we have in our URDF?

What do we write inside an <inertial> tag?

We will create three Xacro macros containing an `<inertial>` tag—one for a box, one for a cylinder, and one for a sphere. Inside a URDF `<inertial>` tag, you will need to provide:

- The mass of the element (in kg).
- The origin for the inertia (in meters and radians).
- The nine elements of the inertia tensor, or matrix (in kg per sqm). Since the matrix is symmetrical, we only need six elements—`ixx`, `ixy`, `ixz`, `iyy`, `iyz`, and `izz` (for example, `ixy` and `iyx` are the same, so we omit the second one).

As we don't have a physical robot for this project, we will arbitrarily decide on a mass property for each link—while, of course, trying to have values that make sense.

Now, how do we compute the inertia matrix? This is usually the hardest part when writing the `<inertial>` tags.

If you are designing your robot with CAD software—for example, with **SolidWorks**—then you can export each property directly from the software and add them to your URDF. As we don't have this software, we will need to make the computation ourselves.

Fortunately, there are some helpful resources on the internet. You can find a list of moments of inertia on Wikipedia: `https://en.wikipedia.org/wiki/List_of_moments_of_inertia`. There, you can also find the moment of inertia for each simple shape that we have, as well as a list of 3D inertia tensors, which are basically the matrices we need for the URDF. One thing to note is that the matrices only have three non-zero components—`ixx`, `iyy`, and `izz`. All the other components are set to 0.

With this information, we can start writing the `<inertial>` tags.

Adding inertial macros for basic shapes

As the inertial macros for basic shapes could be used by any robot, we will add all macros in the `common_properties.xacro` file. This way, if you want to create another URDF for another robot, you can just reuse this Xacro file.

The first macro will be for a box inertia. Now, if you look at the preceding Wikipedia link, things could be a bit confusing, as they use **width**, **depth**, and **height** (*w*, *d*, and *h*). In ROS 2, we have specified the length, width, and height for the *x*, *y*, and *z* dimensions. Which one corresponds to which?

One easy way to write this matrix properly is to realize that one component for one dimension uses the two other dimensions. For example, to compute the *w* component (`ixx` in the matrix), we use *h* (*z*) and *d* (*y*). Only following this can remove a lot of confusion, especially with the different naming conventions.

Here is what we will use (on the left, the Wikipedia value, and on the right, the ROS 2 value):

- **w**: the *x* dimension
- **d**: the *y* dimension
- **h**: the *z* dimension (this is also the axis pointing up)

Here is the `<inertial>` tag for a box. You can add this after the `<material>` tags, inside the `<robot>` tag:

```
<xacro:macro name="box_inertia" params="m x y z o_xyz o_rpy">
    <inertial>
        <mass value="${m}" />
        <origin xyz="${o_xyz}" rpy="${o_rpy}" />
        <inertia ixx="${(m/12) * (z*z + y*y)}" ixy="0" ixz="0"
                 iyy="${(m/12) * (x*x + z*z)}" iyz="0"
                 izz="${(m/12) * (x*x + y*y)}" />
    </inertial>
</xacro:macro>
```

To make it simple, I used x, y, and z directly (instead of *w*, *d*, and *h*), which will make things easier when we need to use the macro in our links, where the box dimensions are defined with x, y, and z. When you develop an API/interface/macro/function, the best practice is to design the interface for the client of the API, not for the developer who's writing the API.

Let's now write the macro for a cylinder. This one is a bit easier. We have two components—radius and height. This will correspond to the radius and length we defined in the URDF:

```
<xacro:macro name="cylinder_inertia" params="m r l o_xyz o_rpy">
    <inertial>
        <mass value="${m}" />
        <origin xyz="${o_xyz}" rpy="${o_rpy}" />
        <inertia ixx="${ (m/12) * (3*r*r + l*l) }" ixy="0" ixz="0"
                 iyy="${ (m/12) * (3*r*r + l*l) }" iyz="0"
                 izz="${ (m/2) * (r*r) }" />
    </inertial>
</xacro:macro>
```

Finally, we can write the macro for a (solid) sphere. This is the easiest, and we only have one component —the sphere radius:

```
<xacro:macro name="sphere_inertia" params="m r o_xyz o_rpy">
    <inertial>
        <mass value="${m}" />
        <origin xyz="${o_xyz}" rpy="${o_rpy}" />
        <inertia ixx="${ (2/5) * m * r * r }" ixy="0" ixz="0"
                 iyy="${ (2/5) * m * r * r }" iyz="0"
                 izz="${ (2/5) * m * r * r }" />
    </inertial>
</xacro:macro>
```

With those three macros, we have everything we need for all the basic URDF shapes.

Including the inertial macros in the links

We can now use those macros to provide the inertial property for each link of the robot.

As the `base_footprint` doesn't represent a physical part (it's what we call a virtual link), it won't have an inertia. For all the other links (base, right wheel, left wheel, and sphere), we will use the inertial macros.

Open the `mobile_base.xacro` file, which is where we will continue with the code.

Now, to add the inertial property for a link, you need to add an `<inertial>` tag inside the `<link>` tag. To add this tag, we will use the macros we previously created.

Let's start with `base_link`. Inside the `<link name="base_link"></link>` tag, and after the `<visual>` tag, add the `box_inertia` macro:

```
<xacro:box_inertia m="5.0" x="${base_length}" y="${base_width}"
z="${base_height}" o_xyz="0 0 ${base_height / 2.0}" o_rpy="0 0 0" />
```

> **Note**
>
> Both the `<visual>` and `<inertial>` tags are children of the `<link>` tag; do not add the `<inertial>` tag inside the `<visual>` tag.

We specify all the parameters required for the macro:

- **Mass**: Here, we decided that the box would be `5.0` kg.
- **Box properties**: The `x`, `y`, and `z` dimensions (we have created the macro so that we can use the ROS 2 axis system convention).
- **Inertia origin**: The inertia is related to the box itself, so if the box has an offset relative to the joint origin, you need to take this offset into account. Basically, you can use the same values you wrote in the visual origin. If you are confused, don't worry—after adding the `<inertial>` tags, we will verify the inertia in RViz. There, you can easily see whether the inertia is correctly placed.

Now, add the inertia for the two wheels. You will add the `cylinder_inertia` macro inside the `wheel_link` macro (it's perfectly fine to use a macro within another macro). Make sure to place it inside the `<link>` tag and after the `<visual>` tag. This inertia macro will apply to both wheels:

```
<xacro:cylinder_inertia m="1.0" r="${wheel_radius}" l="${wheel_
length}" o_xyz="0 0 0" o_rpy="${pi / 2.0} 0 0" />
```

Here are the parameters we specify:

- **Mass**: Set the mass for each wheel to `1.0` kg.
- **Cylinder properties**: The radius and length of the cylinder.
- **Inertia origin**: The visual is centered around the link origin, so we don't need to add any offset. However, to match the image on the Wikipedia page, we have defined the macro for the inertia of a cylinder with the rotation axis as the z axis. The wheels' visual has been shifted by 90° on the x axis so that the rotation axis becomes the y axis. Here, we provide the same rotation for the origin. Basically, once again, you can use the same values you wrote in the visual origin.

Finally, we add the `sphere_inertia` macro for `caster_wheel_link`:

```
<xacro:sphere_inertia m="0.5" r="${wheel_radius / 2.0}"
    o_xyz="0 0 0" o_rpy="0 0 0" />
```

This one is quite easy. We choose `0.5` kg for the mass, and then we provide `radius`, which is the only sphere property. We don't need to specify any offset for the origin.

That's it for the `<inertial>` tags. The main difficulty was to define the macro for the simple shapes (but you only need to do this once), and then to correctly use the macros and provide the correct origin. Let's now verify that we provided the correct dimension values and origins, using RViz.

Validating inertia with RViz

To make sure that the inertial property is correct for each link, you can use RViz. As we already have created a launch file inside the `my_robot_description` package, let's use it to visualize the URDF in RViz.

Don't forget to build and source the workspace before you start RViz. Then, start the launch file:

```
$ ros2 launch my_robot_description display.launch.xml
```

You will see the same view and configuration from the previous chapter. To see the inertia, first disable the visual. On the left menu, open **RobotModel** and uncheck **Visual Enabled**. Then, still inside **RobotModel**, open **Mass Properties** and check the **Inertia** box. You should see something like this:

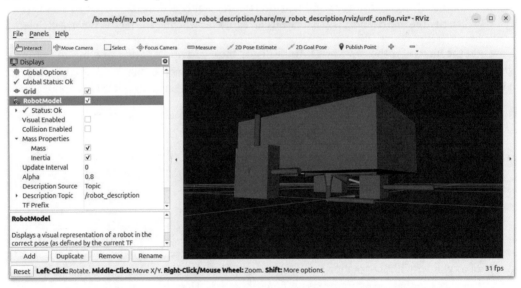

Figure 13.3: Visualizing inertia in RViz

With this view, you can easily spot errors. For example, if the offset for the `base_link` inertia is not right, then the box will not be correctly placed. Another common error will be in the rotation of the wheel inertia. In the preceding figure, you can see that the inertia box is more or less on top of the wheel, correctly orientated. If that's not the case, you know you have to fix the inertia in the code.

Once you are done with the `<inertial>` tags, you can add the `<collision>` tags.

Collision tags

So far, inside each link of our URDF, we have a `<visual>` tag to see the link and an `<inertial>` tag to describe the physical properties for Gazebo.

However, there is something missing. The visual is only for you to visualize the link in RViz or Gazebo. However, Gazebo needs more than this to simulate the robot. You will have to add `<collision>` tags to the links so that Gazebo can compute how two parts collide with each other.

To give you an idea, if you don't have any collision property for your robot, then the robot will fall through the ground and continue falling indefinitely. With a collision property, the robot will *collide* with the ground and, thus, not fall. On top of that, that property will be used to compute collisions between different parts of the robot, or collisions with other robots and elements in an environment. For example, if the robot collides with a wall or an object, it will bump into it and not go through it. Once again, the `<visual>` tag doesn't do any of this.

Before we write the `<collision>` tags for our robot, let's understand how to define a collision.

How to define a collision element

A `<collision>` tag will contain more or less the same thing as a `<visual>` tag: `<geometry>` and `<origin>` tags. You will basically define a shape.

As a general rule, for a collision, you will use a simpler shape than for the visual (if possible). The reason is simple—the more complex the shape, the more computation power will be required to compute the collision between the link and other elements. This could slow down the simulation a lot. Thus, designing simpler shapes is a best practice.

Here are a few more details about defining a collision element:

- If you are using complex Collada files (a few MB) for visuals, use simpler Collada or even STL files for the collisions. You can add those files to the `meshes` folder and include them in the URDF.
- If the shape is close to a basic one (a box, cylinder, or sphere), then you can use a basic shape for the collision. For example, if you design a mobile robot and the base of the robot looks like a box, then you can use a complex Collada or STL file for the visual, only using a box for the collision. For a wheel, you can use a cylinder, sphere, and so on.
- Even when using basic shapes, you could reduce the complexity—for example, by using a box for a collision when the visual is a cylinder or a sphere (however, we will see an exception to this later in this section in order to reduce friction in Gazebo).

To find real examples of this shape simplification, you can simply search on GitHub for existing projects. Find a robot powered by ROS 2, and type `<name of the robot> + description + github` into Google—you can try with the *TurtleBot 3* (or a more recent version) robot. You will usually find a package similar to the one you created in the previous chapter. In this package, search for the URDF or Xacro files, and find the `<collision>` tags. You will often see that the collision uses a simplified STL file or a basic URDF shape.

Now that we have this simplification mindset, let's add the `<collision>` tags inside the URDF.

Adding collision properties for the links

You will add `<collision>` tags inside the `<link>` tags, in the `mobile_base.xacro` file.

Pay attention to where you add those tags. They should be at the same level as the `<visual>` and `<inertial>` tags, not inside them. The order doesn't really matter, but to keep things clean, I usually start with `<visual>`, then `<collision>`, and finally, `<inertial>`.

Here is the `<collision>` tag for `base_link`:

```
<collision>
    <geometry>
        <box size="${base_length} ${base_width} ${base_height}" />
    </geometry>
    <origin xyz="0 0 ${base_height / 2.0}" rpy="0 0 0" />
</collision>
```

A box is already the simplest shape you can have, so we just use another box. Then, we set the same dimensions and origin as for the visual. As you can see, adding the collision for this link basically means copying and pasting the content of the `<visual>` tag (except for `<material>`).

Now, for the two wheels, the situation is a bit unique. They are cylinders, so it's a quite basic shape. We could keep the same, or even use a box, which is much simpler.

However, those wheels will be in direct contact with the floor in Gazebo. Using a box is not a good idea—imagine a car with square wheels; it's probably not going to work well.

We could keep a cylinder, which is the second most simple shape, but if you do this, you might experience some unwanted friction later on in Gazebo, which will result in the robot not moving exactly how we want. For wheels that you control and that touch the floor, it's best to use a sphere for the collision, as this reduces the number of contact points with the ground (basically, just one contact point) and, thus, reduces the unwanted friction.

> **Note**
> This is more of a Gazebo *hack* than a real logical choice. Let's be clear here—there is no way you can come up with this conclusion at this point; it's something you have to experience in Gazebo and fix later. I'm not a big fan of such hacks, but sometimes, you don't have a choice. Also, I'm giving you the correct solution beforehand to not make this chapter too long.

Let's add a sphere collision for our wheels, inside the `wheel_link` macro:

```
<collision>
    <geometry>
        <sphere radius="${wheel_radius}" />
    </geometry>
    <origin xyz="0 0 0" rpy="0 0 0" />
</collision>
```

We just use the wheel radius property; there is no need for the wheel length here. And since it's a sphere, there's no need to add any rotation for the origin.

Finally, we add the collision for the caster wheel. As stated before, we could simplify the sphere with a box, but as the caster wheel is in direct contact with the ground, we don't want to add unwanted friction. Thus, we use the same sphere for the visual and the collision:

```
<collision>
    <geometry>
        <sphere radius="${wheel_radius / 2.0}" />
    </geometry>
    <origin xyz="0 0 0" rpy="0 0 0" />
</collision>
```

That's it for the `<collision>` tags. The URDF for our mobile robot is now complete.

Validating collision with RViz

As with everything we have written in the URDF, you can also visualize the collision properties in RViz.

Build your workspace, source the environment, and start the `display.launch.xml` launch file again.

Open the **RobotModel** menu and uncheck the **Visual Enabled** box. Then, check the **Collision Enabled** box.

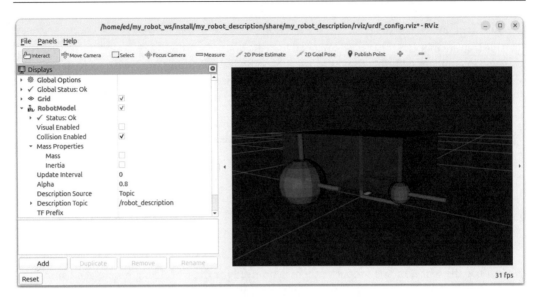

Figure 13.4: Visualizing collision in RViz

There, you can see whether the collision properties are correct. In the **Links** menu (inside **RobotModel**), you can enable only some links if necessary, giving you a more precise view.

If you see that some collision elements are not correctly placed, or if they are too big or too small, you can then go back to your URDF and fix them.

As you can see in the preceding figure, the collision view for the robot is almost the same as the visual. The difference is with the wheels, which are now spheres. You might also wonder, if the wheels are spheres, a part of them is inside the box, and thus they will collide with the box. That's true, but this collision between adjacent links will not be taken into account by Gazebo.

Now that you've added both the `<inertial>` and `<collision>` tags, we can go to the next step and spawn the robot in Gazebo.

Spawning the robot in Gazebo

The first step of adapting the URDF for Gazebo was crucial, as without this, the robot either wouldn't appear on Gazebo or behave incorrectly.

Now that the URDF is done (and has been verified in RViz), we can spawn the robot in Gazebo. In this section, you will see what commands you need to run, and then we will create another package with a launch file to start everything.

Here is what we will do:

1. Run the `robot_state_publisher` node with the URDF as a parameter.
2. Start the Gazebo simulator.
3. Spawn the robot in Gazebo.

Let's start this with the terminal commands.

Spawning the robot from the terminal

As usual, before creating a launch file, we will run each command one by one in different terminals, enabling us to clearly understand what we need to run, with all the necessary details.

Open three terminals to start all the commands.

The first thing to start is the `robot_state_publisher` node. This is what we did in *Chapter 12*, and this is also usually the first thing you will start in any ROS 2 application.

In Terminal 1, run the following command:

```
$ ros2 run robot_state_publisher robot_state_publisher --ros-args -p
robot_description:="$(xacro /home/<user>/my_robot_ws/src/my_robot_
description/urdf/my_robot.urdf.xacro)"
```

This is the same as in the previous chapter. We pass the URDF with the `robot_description` parameter.

After executing this command, the `robot_state_publisher` node starts and does three things—subscribes to `/joint_states`, publishes on `/tf`, and also publishes the URDF on `/robot_description`. You can verify this with `rqt_graph` if needed (in this case, make sure to uncheck the **Dead sinks** and **Leaf topics** boxes).

In a second terminal, we start Gazebo. In fact, you could start Gazebo before `robot_state_publisher`; the order doesn't matter for those first two steps.

In Terminal 2, run the following command:

```
$ ros2 launch ros_gz_sim gz_sim.launch.py gz_args:="empty.sdf -r"
```

With this, we start an empty world in Gazebo. By default, the time is stopped when you launch Gazebo. We add the `-r` option to start the time directly so that we don't have to click the play button.

Finally, once you have done the first two steps, you can spawn the robot in Gazebo. For this, we will use the `create` executable from the `ros_gz_sim` package. This will spawn a robot in Gazebo, using the URDF of the robot, which we can pass as a topic with the `-topic` option. As the `robot_state_publisher` node publishes the URDF on the `robot_description` topic, we can use this topic.

In Terminal 3, run the following command:

```
$ ros2 run ros_gz_sim create -topic robot_description
```

After running this command, you should see the robot in Gazebo:

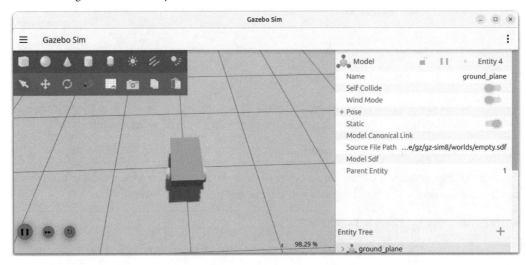

Figure 13.5: The robot spawned in Gazebo

If you get any error in any terminal, it probably means that your URDF is not correct. In this case, go back to the Xacro files and double-check everything.

> **Note**
> To see the importance of the `<inertial>` and `<collision>` tags, go back to the URDF, comment the inertia for the wheels, and then run the commands again. You will see that the wheels do not appear in Gazebo. Then, put the inertia back and comment the collision for the wheels. This time, the wheels will appear, but as they do not collide with the ground, you will see them going into the ground.

Now that we know which commands to run, let's write a launch file.

Spawning the robot from a launch file

We are now going to write a launch file to start those three commands. This will be a good base on which we can add more things later on.

Let's start by creating a new package in our workspace.

Creating a _bringup package

If you remember what we did in *Chapter 9*, the best practice is to create a dedicated package for launch files and configuration files (we don't have any configuration files yet, but we will add one later in this chapter).

We made an exception for the `display.launch.xml` file, which we placed inside the `my_robot_description` package. As explained in *Chapter 12*, this launch file is only used during development to visualize the URDF. Thus, it makes sense to have the launch file in the same package as the URDF. Here, and for any future launch and configuration files in our application, we will use a new dedicated package.

Following the naming convention for such a package, we will start with the name of the robot or application, adding the `_bringup` suffix. Thus, we will create the `my_robot_bringup` package.

> **Note**
>
> Be careful not to confuse this package with the `my_robot_bringup` package in the `ros2_ws` we created in *Part 2* of the book. Here, in *Part 3*, we are using another workspace, named `my_robot_ws`, so the `my_robot_bringup` package is a completely different one.

Let's create this package, remove the unnecessary folders, and add a `launch` folder. We will also add a `config` folder, which we will use later in the chapter:

```
$ cd ~/my_robot_ws/src/
$ ros2 pkg create my_robot_bringup --build-type ament_cmake
$ cd my_robot_bringup/
$ rm -r include/ src/
$ mkdir launch config
```

Now, in the `CMakeLists.txt` file of the `my_robot_bringup` package, add the instruction to install the `launch` and `config` folders:

```
install(
  DIRECTORY launch config
  DESTINATION share/${PROJECT_NAME}/
)
```

The package is correctly set up, so we can now add and install files.

Writing the launch file

Let's create, write, and install the launch file to spawn the robot in Gazebo.

First, create a new file inside the launch folder. As this launch file will be the main one, let's simply use the name of the robot (or robotics application), my_robot.launch.xml:

```
$ cd ~/my_robot_ws/src/my_robot_bringup/launch/
$ touch my_robot.launch.xml
```

Open the file and write the minimal code for an XML launch file:

```xml
<launch>
</launch>
```

Then, inside this <launch> tag, let's add everything we need, step by step.

The beginning of the launch file will be very similar to the display.launch.xml file that we wrote in the previous chapter, so we can basically copy and paste a few parts. We start by adding a variable for the path to the URDF file:

```xml
<let name="urdf_path" value="$(find-pkg-share my_robot_description)/urdf/my_robot.urdf.xacro" />
```

Now, we can start the robot_state_publisher node:

```xml
<node pkg="robot_state_publisher" exec="robot_state_publisher">
    <param name="robot_description"
           value="$(command 'xacro $(var urdf_path)')" />
</node>
```

Then, we start Gazebo with an empty world, and we also use the -r option to start the time automatically:

```xml
<include
    file="$(find-pkg-share ros_gz_sim)/launch/gz_sim.launch.py">
    <arg name="gz_args" value="empty.sdf -r" />
</include>
```

Finally, we spawn the robot in Gazebo:

```xml
<node pkg="ros_gz_sim" exec="create" args="-topic robot_description"
/>
```

That's it for the launch file for now. Later, we will add more things and also start RViz to visualize the TFs. For now, we just want to see the robot in Gazebo.

There is one thing we can do to make things a bit cleaner—as we are using files, nodes, and launch files from other packages, let's add a dependency to them in the `package.xml` file of the `my_robot_bringup` package. After the `<buildtool_depend>` line, add these lines:

```
<exec_depend>my_robot_description</exec_depend>
<exec_depend>robot_state_publisher</exec_depend>
<exec_depend>ros_gz_sim</exec_depend>
```

We use the less strict `<exec_depend>` tag instead of `<depend>`, as we only require those dependencies to run the launch file, not to compile any code. With this, if, for example, you didn't install the `ros_gz_sim` package and you try to build the `my_robot_bringup` package, you will get an error when running `colcon build`, and then you can fix things right away. Without those lines, the build would work, but you would then get an error when starting the launch file, which can be a big problem, especially in a production environment. So, the best practice is to specify all the dependencies you need in the `package.xml` file.

Now, save all the files, build the workspace, source the environment, and start the launch file (make sure that Gazebo is not running in another terminal before you do this):

```
$ ros2 launch my_robot_bringup my_robot.launch.xml
```

You should get the same result as when we ran all three commands in the terminal.

Controlling the robot in Gazebo

Our mobile robot is now simulated in Gazebo with physics properties. Now what? The robot is not doing anything. We will finish this chapter by adding control plugins so that we can simulate the hardware of the robot and do the following:

- Send commands to make the robot move in Gazebo, just as if it were in the real world
- Read all necessary joint states from the robot to get all the TFs in our ROS 2 application

Before we start discussing Gazebo systems and bridges, let's dive a little bit deeper and understand what's missing and what we need to add.

What do we need to do?

When you start the `my_robot.launch.xml` launch file, you see the robot in Gazebo. However, we don't have any way to control it. In a terminal, if you list all nodes, topics, services, or even actions, you won't find anything we can use.

Also, after starting the launch file, if you print the TF tree, you won't see the TF for the right or left wheel. You can observe the same thing with RViz—to make things simple, you can start RViz using the previous configuration we saved:

```
$ ros2 run rviz2 rviz2 -d ~/my_robot_ws/src/my_robot_description/rviz/urdf_config.rviz
```

You should see some errors in **RobotModel**, saying **No transform from [left_wheel_link]** and **No transform from [right_wheel_link]**.

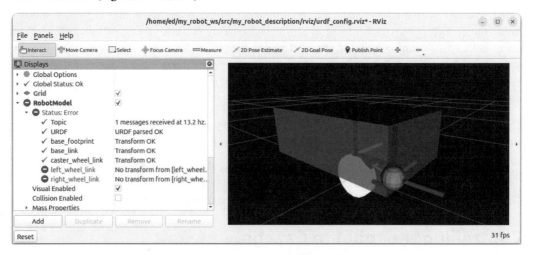

Figure 13.6: TF errors in RViz after spawning the robot in Gazebo

This lack of TF is because nobody is publishing on the `/joint_states` topic. In *Chapter 12*, when we were just visualizing the robot model, we used a fake joint state publisher. We won't do this here.

So, what do we need to do?

For a real robot, you would create a hardware driver to be able to control the wheels. This driver would expose a topic/service/action interface so that you could make the robot move. Then, you would read position/velocity data from encoders and publish this data on the `/joint_states` topic. With this, the loop is closed.

For a Gazebo simulation, we will do the same thing but, of course, without hardware. We will use Gazebo plugins (also called systems) to simulate the control of the robot and get the joint states. Then, we will configure a *bridge* to make those plugins communicate with our ROS 2 application.

Let's start with the Gazebo systems.

Adding Gazebo systems

A Gazebo system is basically the simulation of a hardware component. You could have a system simulating a camera and publishing images, another one monitoring a battery state, and so on. For this book, we will use two systems—one to control a differential drive robot (two parallel wheels), and one to publish the joint states.

Now, the good news is that there are a lot of existing systems already available to use, including the two that we need.

The bad news is that documentation for those systems is almost non-existent (at the time of writing), and you will have to dive into the code itself to find what to include in your own code. Don't worry about that—we will do this process step by step, and it is replicable for any other system that you use.

For Gazebo Harmonic and ROS 2 Jazzy, you can find all available Gazebo systems here on GitHub: https://github.com/gazebosim/gz-sim/tree/gz-sim8/src/systems (for other Gazebo versions, you might have to use a different branch).

> **Note**
>
> If there was not enough confusion already, on the internet you will often see the term *plugin* or *system*; they both refer to the same thing. Even if the word *system* should be preferred, in practice it's not clear which one to use; for example, to include a system in our code, we will need to use a `<plugin>` tag. So, in this section, I will have to use both terms.

Now, where are we going to add our systems for the robot we want to simulate? We will do this in the URDF.

Xacro file for Gazebo

The Gazebo systems for our robot will be specified in the URDF. So, we need to return to the `my_robot_description` package.

Our URDF is now split into three files: one with common properties, one with the description of the robot (links and joints), and one to include the other two.

To add the Gazebo systems, we will create yet another Xacro file, dedicated to all Gazebo-related stuff. By separating this file from the other ones, we make things cleaner. If, later on, you want to use the URDF without Gazebo, you only need to remove the inclusion of the Gazebo file.

In the `urdf` folder of your `my_robot_description` package, add a fourth file, named `mobile_base_gazebo.xacro`.

Open the file and add the minimal Xacro code:

```xml
<?xml version="1.0"?>
<robot xmlns:xacro="http://www.ros.org/wiki/xacro">
</robot>
```

Now, in `my_robot.urdf.xacro`, include the file after the two other ones:

```xml
<xacro:include filename="$(find my_robot_description)/urdf/mobile_base_gazebo.xacro" />
```

The Xacro file is ready, and we can now add the systems.

Differential drive controller

The first system we will add is a **differential drive controller**. By differential drive, we mean a robot controlled by two wheels, one on each side of the robot.

If you browse the available systems (the link is provided on the preceding page), you can find a `diff_drive` folder—usually in ROS, we use **diff drive** as an abbreviation of **differential drive**.

In this folder, you will see a `DiffDrive.hh` file. Open this file, and there, near the beginning, you will find the XML tags related to the system (it's possible that some tags will be missing here; for some systems, you might have to read the complete source code to find all available tags).

Here is how to add the system to our Xacro file (`mobile_base_gazebo.xacro`):

```xml
<gazebo>
    <plugin
        filename="gz-sim-diff-drive-system"
        name="gz::sim::systems::DiffDrive">
        <left_joint>base_left_wheel_joint</left_joint>
        <right_joint>base_right_wheel_joint</right_joint>
        <frame_id>odom</frame_id>
        <child_frame_id>base_footprint</child_frame_id>
        <wheel_separation>0.45</wheel_separation>
        <wheel_radius>0.1</wheel_radius>
    </plugin>
</gazebo>
```

We start with a `<gazebo>` tag. Everything related to Gazebo will be in such tags. Then, we include the system with a `<plugin>` tag. We also need to specify the filename and name for the system.

> **Note**
>
> Usually, the filename and name will follow this syntax:
>
> - **Filename**: `gz-sim-<name-with-dashes>-system`
>
> - **Name**: `gz::sim::systems::<UpperCamelCaseName>` (you can also find the name at the bottom of the `.cc` file of the system)

Here is a bit more information about the different parameters for this diff drive system:

- `left_joint` and `right_joint`: You need to provide the exact name of the joints you have defined for the wheels in the URDF.
- `frame_id`: As the robot moves, we will keep track of where it is relative to its starting position. This starting position will be called `odom` (short for **odometry**).
- `child_frame_id`: We write `base_footprint`, as it is the root link for our robot and the one we want to use for odometry tracking.
- `wheel_separation`: We can compute that from the URDF. The base width is 0.4, and the origin for each wheel is centered on the wheel. As each wheel length is 0.05, we need to add 0.4 + 0.025 + 0.025, which makes `0.45`.
- `wheel_radius`: We get this value from the URDF, which is defined as `0.1`.
- Acceleration and velocity min and max: Optionally, you can set some limits. This can be a good idea so that the controller doesn't accept a command that would make the robot move too fast and, potentially, become dangerous to itself or an environment. For the values, once again, you should use the metric system and radians for angles.

That's it for the diff drive system. Now, in addition, there is one setting we need to add for the caster wheel. If you remember, the caster wheel is a passive joint, so we defined it as a fixed sphere.

As the wheels turn and the robot moves, there will be some friction between the ground and the caster wheel. You won't see the friction that much in Gazebo, but it will slow down the robot a bit, and later on, if you visualize the robot in RViz, you won't have the same result.

So, we will reduce the friction for the caster wheel. You can add this code just before the code for the diff drive system:

```
<gazebo reference="caster_wheel_link">
    <mu1 value="0.1" />
    <mu2 value="0.1" />
</gazebo>
```

There are two parameters, `mu1` and `mu2`, that you can set to have more control over the friction. I have chosen the value `0.1`; later, you could reduce this value even more.

Joint state publisher

We have added a system to control the wheels, but before we test it, let's finish the Xacro file and add the second system we need. The diff drive system alone won't publish the joint states for the wheels; we need to add a joint state publisher system.

Go back to the systems page on GitHub, and you will find a `joint_state_publisher` folder. In this folder, you can get the *documentation* for the XML tags in the `JointStatePublisher.hh` file.

Let's add the system to the Xacro file, after the previous one:

```xml
<gazebo>
    <plugin
        filename="gz-sim-joint-state-publisher-system"
        name="gz::sim::systems::JointStatePublisher">
    </plugin>
</gazebo>
```

The joint state publisher system is easier to set up. Also, we don't specify any `<joint_name>` tag here to publish all available joint states. If your robotics system contains a lot of joints, it could be useful to only specify the joints you want to use.

Our `mobile_base_gazebo.xacro` file is now complete, and we won't need to modify anything else in the URDF. We can spawn the robot in Gazebo again and see how it interacts with those systems.

Bridging Gazebo and ROS 2 communications

The last thing we need to do, for this simulation to be complete, is to bridge Gazebo and ROS 2 communications.

Let's first understand what's missing.

What topics do we need to bridge?

If you remember, we talked about this at the beginning of the chapter. Gazebo uses topics and services, but those are independent from ROS 2. Thus, the systems we have just added will work, but they will only have a Gazebo interface.

You can verify this by starting the `my_robot.launch.xml` file again—make sure to compile and source the workspace beforehand so that you get the updated URDF.

Then, in another terminal, list all Gazebo topics. The list will contain quite a lot of things; here, I only include the ones that we will use:

```
$ gz topic -l
/model/my_robot/tf
/world/empty/model/my_robot/joint_state
/model/my_robot/cmd_vel
```

The first topic that ends with /tf will contain the TF from the odom frame to base_footprint. The one with /joint_state will contain the joint states for both wheels, and the topic with /cmd_vel will be used to send a velocity command to the robot.

However, if you check the ROS 2 topics with ros2 topic list, you won't see the /cmd_vel topic. You will have /joint_states and /tf, but only because the robot_state_publisher node creates a subscriber and publisher for those topics. Nothing is published; you can verify this with ros2 topic echo <topic>.

Thus, from the ROS 2 side, we can't communicate with Gazebo. We will need to create a bridge between ROS 2 and Gazebo using the ros_gz_bridge package (see *Figure 13.2* at the beginning of the chapter).

To do that, we will run the parameter_bridge node from the ros_gz_bridge package, with a configuration for the interfaces that we want to bridge.

Adding a configuration file to bridge topics

Let's start with the configuration file. In the my_robot_bringup package, inside the config folder (that we already created before), create a new file named gazebo_bridge.yaml.

Open this file to write the configuration. Here is the first bridge we will create:

```
- ros_topic_name: "/cmd_vel"
  gz_topic_name: "/model/my_robot/cmd_vel"
  ros_type_name: "geometry_msgs/msg/Twist"
  gz_type_name: "gz.msgs.Twist"
  direction: ROS_TO_GZ
```

Here are the different fields that we will use:

- ros_topic_name: The topic name on the ROS 2 side. Either you choose the topic name (/cmd_vel doesn't exist yet, so we create it) or you make it match with an existing one (for the next one, we will have to specify exactly /joint_states).
- gz_topic_name: The topic name on the Gazebo size. We found it with gz topic -l.
- ros_type_name: The topic interface for ROS 2.

- `gz_type_name`: The topic interface for Gazebo. You can find it with `gz topic -i -t <topic>`.
- `direction`: Either `ROS_TO_GZ`, `GZ_TO_ROS`, or `BIDIRECTIONAL`. For example, `/cmd_vel` is a topic that we publish in ROS 2 and subscribe in Gazebo, so we use `ROS_TO_GZ`. For `/joint_states`, we publish in Gazebo and subscribe in ROS 2, so that will be `GZ_TO_ROS`. You can use `BIDIRECTIONAL` if you want to have publishers and subscribers on both sides of the same topic.

As you can see, we need to provide the topic name and interface on both sides and specify which direction to use for the communication. With this, the `ros_gz_bridge` will create the connection.

With this first bridge, we will be able to send commands to the robot to make it move with the diff drive system. Let's now add the configuration for the `/joint_states` topic (published by the joint state publisher system):

```
- ros_topic_name: "/joint_states"
  gz_topic_name: "/world/empty/model/my_robot/joint_state"
  ros_type_name: "sensor_msgs/msg/JointState"
  gz_type_name: "gz.msgs.Model"
  direction: GZ_TO_ROS
```

That will allow us to get all joint states for the robot and, thus, see the wheel TFs in RViz. Finally, to get the `odom` to `base_footprint` TF (published by the diff drive system), we also add this bridge:

```
- ros_topic_name: "/tf"
  gz_topic_name: "/model/my_robot/tf"
  ros_type_name: "tf2_msgs/msg/TFMessage"
  gz_type_name: "gz.msgs.Pose_V"
  direction: GZ_TO_ROS
```

The configuration file is complete. As we have already added the instruction to install it in `CMakeLists.txt`, there is no need to do anything else.

Starting the Gazebo bridge with the configuration

We can now add a new node to our `my_robot.launch.xml` file to start the bridge, using the YAML configuration file we've just created.

First, at the beginning of the file, let's add a new variable to find the path for the configuration file:

```
<let name="gazebo_config_path" value="$(find-pkg-share my_robot_bringup)/config/gazebo_bridge.yaml" />
```

Then, after you spawn the robot in Gazebo with the `create` executable from `ros_gz_sim`, start the Gazebo bridge. You will need to pass the configuration file inside a `config_file` parameter:

```
<node pkg="ros_gz_bridge" exec="parameter_bridge">
    <param name="config_file"
           value="$(var gazebo_config_path)" />
</node>
```

As we use the `ros_gz_bridge` package inside `my_robot_bringup`, we will also add a new dependency inside the `package.xml` file:

```
<exec_depend>ros_gz_bridge</exec_depend>
```

The Gazebo bridge is now correctly configured. When you start your application, ROS 2 and Gazebo will be able to communicate with each other.

Testing the robot

In this final section, we will make sure that everything works by testing the behavior of the robot, and also by visualizing the robot and TFs in RViz.

Save all files, build and source the workspace, and start the `my_robot.launch.xml` file again.

In another terminal, list all the topics, and you will see the `/cmd_vel` topic that we configured previously. The interface for this topic is the same one we used for Turtlesim in *Part 2* of the book, so you should be familiar with it. Send a velocity command from the terminal:

```
$ ros2 topic pub /cmd_vel geometry_msgs/msg/Twist "{linear: {x: 0.5}}"
```

The robot should start moving in Gazebo (to stop, send the same command with `{x: 0.0}`). If you see the robot moving, it means that the bridge is correctly configured, as the ROS 2 topic can reach the Gazebo system. It also means that the diff drive system works.

To achieve a better way to control the robot and make more tests, you can run this node instead:

```
$ ros2 run teleop_twist_keyboard teleop_twist_keyboard
```

This will listen to your keyboard and publish to the `/cmd_vel` topic (if you use a different name for the topic, simply add a remapping with `--ros-args -r`).

Now, we have validated that the robot can move in Gazebo when we send a command.

To check the TFs, you can do the following:

- Subscribe to the `/joint_states` topic and see the states for both the right and left wheels
- Subscribe to the `/tf` topic and see all published TFs

- Print the TF tree (`ros2 run tf2_tools view_frames`), which should contain all the TFs, including the two TFs for the wheels and an additional one between the `odom` and `base_footprint`

> **Note**
> If something does not work or if some topic data is missing, then either one of the systems or bridges is not correctly configured. To solve this, first check the topics on the Gazebo side (the `gz topic` command line). If you see the correct data, then the bridge is wrong; if you don't, start with the system.

So, we can control the robot and get the correct TFs in our ROS 2 application. Finally, let's start RViz. You can use the command line, but you can also add RViz directly inside the launch file if you want to. In this case, we will first create a variable to find the RViz config path:

```
<let name="rviz_config_path" value="$(find-pkg-share my_robot_description)/rviz/urdf_config.rviz" />
```

We will use the file we previously created in `my_robot_description`. You could also create a new RViz configuration file and install it in `my_robot_bringup`. Then, we start RViz after all the other nodes:

```
<node pkg="rviz2" exec="rviz2" args="-d $(var rviz_config_path)" />
```

With this, when you start the launch file, you will have both Gazebo and RViz. The TF errors that we previously got in RViz (see *Figure 13.6*) should not be there anymore.

One thing you can do is to select `odom` as the fixed frame, inside **Global Options**. With this setting, when the robot moves in Gazebo, you will also see it moving from its starting position in RViz.

Our application is now finished. The Gazebo systems correctly work and can communicate with the ROS 2 side. The loop is closed.

Summary

In this chapter, you learned how to simulate your robot in Gazebo.

You first discovered how Gazebo works. Gazebo is a 3D simulation tool that can simulate gravity and the physical properties of your robot in the environment—unlike RViz, which is only a visualization tool, helpful for developing and debugging.

Then, you followed the process to simulate a robot in Gazebo. Here is a recap of the steps:

1. Before you even get started, make sure you have a URDF that properly describes all the links and joints of your robot (this is what we did in the previous chapters).
2. Adapt the URDF for Gazebo by adding `<inertial>` and `<collision>` tags for each link. You can use RViz to visualize those properties and make sure they are correct.
3. Spawn the robot in Gazebo. To do this, you first start the Gazebo simulator and the `robot_state_publisher` node. Then, you can spawn the robot.
4. Control the robot with plugins (i.e., systems). To use a system, you will add a `<plugin>` tag to your URDF. Then, to be able to connect the Gazebo systems with ROS 2, you can use the `ros_gz_bridge` package and provide the bridge configuration in a YAML file.

All along the way, we organized the application into two packages:

- `my_robot_description`: This contains the URDF, including the links, joints, inertial and collision properties, and Gazebo systems
- `my_robot_bringup`: This contains the launch file to start the application and the YAML configuration file for the Gazebo bridge

The project we started in *Part 3* is now complete. You have a fully working 3D simulation of a robot, and you can apply the whole process (not only from this chapter but also from all the previous ones) to any other custom robot that you create.

Now, of course, that's not the end; there are more things you may want to do with your robot and ROS 2. In the next chapter, we will conclude the book, and we will provide additional resources and tips for you to go further.

14
Going Further – What To Do Next

You have now finished the book—congratulations! Learning ROS 2 is quite a challenge, and you've made a big step.

To recap, this is what you learned:

- *Part 1:* You cleared up some misconceptions, installed ROS 2, and discovered some of the main concepts through experimentation. This set you up for the rest of the book.
- *Part 2:* This is where you learned about the most important ROS 2 concepts: how to write nodes and communicate with topics, services, and actions, and also how to make your application more dynamic with parameters and launch files.
- *Part 3:* You built a simulated robot, and while practicing the core concepts, you learned about TF, URDF, and Gazebo. TFs are the backbone of almost every ROS 2 application.

With this, you have a solid foundation you can use for any other ROS 2 project. Now, I don't want to leave you there and just say that's it. ROS 2 contains a lot more things, and robotics in general is much broader than just ROS 2. Thus, to help you have a better idea for the future, in this last small chapter, I will give you some recommendations on what to do next.

What to do next is not the same for everyone. I will first attempt to provide a general roadmap, and then explore different nuances and details that will help you choose what to learn, depending on what you want to do. I will also share some extra resources that you can use to learn more about ROS 2.

By the end of this chapter, you will have a better idea of what to do next, depending on your project, job, or learning objectives.

In this chapter, we will cover the following topics:

- ROS 2 roadmap – exploration phase
- Learning for a specific goal

ROS 2 roadmap – exploration phase

When learning a technological topic, I would say the usual pattern is as follows:

1. **Discovery phase**: You start with the basics that everybody should learn, get a broad understanding of the technology, and learn how to use it.
2. **Exploration phase**: Once you have the basics, you try to explore different applications, projects, and topics related to the technology. This will make you connect lots of dots and will give you an even better understanding of the global picture. You will also become better technically.
3. **Specialization phase**: It's impossible to be an expert on everything. At some point, to be able to dive deep into a project, to get a job, or to build a career, you will need to specialize in one particular field. After exploring lots of topics, you will have a better idea of what you want to do, or what's most in demand. You can then focus your attention and specialize.

With this book, you have covered the first step. You have learned the basics that you absolutely need and will, for sure, use in almost all your future projects.

What we will cover now is the exploration phase, which comes right after that. In this section, I will attempt to provide you with a ROS 2 roadmap. This roadmap is a (non-exhaustive) list of what you can learn next, in no particular order.

Before we get started, note that I don't recommend following the roadmap exactly as it is. The best way to learn is through projects. So, in this section, we have a skill list, and in the next section, we will explore different project/job examples to see how to pick the skills you need to learn.

Also, this is my own version; not every ROS expert would necessarily agree with me. If you find something that works for you, that's completely fine. The goal here is to make progress.

I didn't include links for this section because a simple Google search with the provided keywords will do. You will mostly find resources in the official documentation, independent tutorials and YouTube videos, GitHub projects, and questions asked in forums.

Common stacks and frameworks

After learning the core programming basics and concepts, such as TF and URDF, a very common next step is to learn about some of the existing ROS 2 **stacks** and **frameworks**, and also learn how to create interfaces between ROS 2 and hardware components.

> **Note**
> You will often see the terms *stack*, *framework*, and other variations. They usually mean the same thing. Basically, they are collections of packages that focus on solving a specific problem.

Among those stacks/frameworks, you can find **Navigation 2** (for **mobile robots**), **MoveIt 2** (for **robotic arms** and grippers), and **ros2_control** (for **hardware control**).

We will be talking about them because they are used in a lot of applications, and knowing about them is very likely to be beneficial for you. Let's start with the hardware interface.

Hardware interface (and ros2_control)

In the end, a robotics developer creates software in order to control pieces of hardware. You make motors move, you read data from sensors, and you add some algorithms in the middle to create a robotics system that does something useful.

The hardware interfacing part is crucial, and unless you only work with simulation, you will have to work with hardware. Thus, I recommend that you get more familiar with how to write a hardware driver (not specifically related to ROS) and how to interface this hardware driver with ROS 2.

You can first try to create your own driver for a simple piece of hardware (anything: a motor, camera, or any other sensor). Once you can control your hardware with Python or C++, include your driver inside a ROS 2 node, and add topics/services/parameters to create a bridge between the driver and ROS 2.

You can also find existing hardware components that have a ROS 2 driver, and have a look at their interface and code (often available on GitHub).

Understanding how to interact with hardware is a big step in your ROS 2 learning.

Then, once you get that, there is a great framework that allows you to make robust interfaces between your ROS 2 application and your hardware drivers. It's called ros2_control. This framework is used in most robots powered by ROS 2. Once you understand how it works, you can set up new robots with hardware connections very fast.

Warning: learning ros2_control is not easy, and the documentation is not what I would call beginner friendly. It also requires you to write code in C++ and to know more about advanced ROS 2 concepts such as lifecycle nodes and components. I don't recommend you to learn ros2_control directly after this book, especially if that was the beginning of your ROS 2 journey.

What I recommend is that you first get more comfortable with creating basic hardware interfaces for ROS 2. Then, as you progress throughout your projects, you will learn more about advanced ROS 2 concepts, and you will get to the point where you can tackle ros2_control.

Navigation 2 stack

The **Navigation 2** stack, also known as **Nav2**, is very popular for a reason: most robots using ROS 2 are mobile robots, and what do you usually do with a mobile robot? You make it navigate autonomously in a physical environment.

Now, with the ROS 2 basics you got from this book, how do you make that happen? There is a big gap between writing nodes and URDFs and making a robot navigate using path-planning algorithms. If you were to implement this by yourself, it would take you a lot of time and effort.

And if you remember, in the introduction of this book, we talked about how most of the time spent on robotics used to be about reinventing the wheel. We don't want to do that.

Fortunately, you can use the Nav2 stack. With this stack, you can easily create a map of the environment with **Simultaneous Localization And Mapping** (**SLAM**), and then use this map to make your robot navigate from one place to another, while avoiding obstacles.

Of course, the Nav2 stack comes with its own challenges but, once you understand it, you can set up a new autonomous mobile robot in no time, and you can easily work with the hundreds of existing robotics projects using this stack.

So, I strongly recommend you learn a bit about Nav2. Even if you don't plan to work with mobile robots, you can get a basic understanding in just a few hours. Going deeper and actually adapting a robot for Nav2 will take longer, but for now, just get the basics.

MoveIt 2

So, there is a stack for mobile robots, and guess what? There is also a stack for robotic arms.

> **Note**
> The two most common robots you will encounter in ROS 2 projects and jobs are mobile robots and robotic arms. Then come drones, but those are more niche and less supported. It doesn't mean that you can't find good packages to help you with drones, but it will for sure be more challenging. Then, even more niche (and less supported), you can find boats, spider robots, submarines, and so on.

If you are working with a robotic arm (let's say with 5, 6, or 7 axes), you will need to find a way to compute positions and trajectories for this arm. How do you make the arm reach a certain point with a defined orientation, or pick up an object and place it somewhere else?

This can become quite a challenge, especially when you learn more about inverse kinematics, motion planning, and when you try to make all the joints of the robot move and arrive at the same time, without colliding with anything, while having continuous positions, velocities, and accelerations.

MoveIt 2 will do the motion planning for you, for a robotic arm, or even a system with several robotic arms. It also has functionalities for grasping.

You will need to do some configuration for your robot (starting from the URDF), and then you can use some Python or C++ APIs directly in your nodes to send commands to control the robot.

I recommend you get at least a basic introduction to MoveIt 2, even if you won't use robotic arms. You can set up a basic project in a few hours and see the main functionalities.

With Nav2 and MoveIt 2, you can cover a lot of ground. I don't know the exact percentage of robots that use either of those stacks, but it's definitely more than half of all ROS 2 robots. Nav2 and MoveIt 2 also both have integrations with ros2_control, which you can explore when you learn more about ros2_control.

Let's now explore more topics related to ROS 2.

More exploration topics

On top of common stacks and frameworks, there are many additional things you could learn or improve. Here, we will explore some of them, starting with the basics. As stated in this section's introduction, for any topic, type the relevant keyword into Google—maybe followed by `tutorial`—and you will find what you need (as explained in previous chapters, for more advanced ROS 2 concepts, documentation might be much more scarce). Once again, there is no particular order you should follow for this list of skills. Later in this chapter, we will see, with a few project and job examples, what to learn, depending on your goals.

Coming back to the basics

Before jumping to advanced ROS 2 concepts, you should make sure that you have the basics right. Maybe when reading this book and doing the exercises, using all the terminal commands was a challenge for you. Or maybe the Python code was OK, but writing with OOP is not something you are used to.

To give you an example, in my courses or workshops, I often see people struggling with typing the right command in the terminal using auto-completion. This is a basic skill that will make you work five times faster with Linux, but if you don't know how to do it properly, your whole ROS 2 learning experience will be slowed down so much. You can't overlook the basics.

At this point, it could be beneficial to improve your skills in the following:

- **Linux**: Especially the command line, and learning more about the environment. Understanding what the `.bashrc` does, navigating to a package installation folder, getting remote access with SSH: these are all examples of what you need to know in order not to be stuck in the future. You don't need to be a Linux expert, but spending a few hours to improve the basics can't hurt.
- **Python**: I guess at this point you should be quite comfortable with Python, but if you had some challenges because of Python (and not ROS 2), then it could be helpful to review some basics, especially with classes and OOP.

- **C++**: If you only looked at the Python examples, now is a good time to try to do the same with C++. Not only will it make you review the concepts one more time, but as you progress with ROS 2, you will realize that a lot of code is only written in C++, especially for hardware control and algorithms that require a lot of computation power. If you want to become a great robotics developer, you will need C++.

Now, once you get those basics right, what other core ROS 2 concepts could you learn?

More advanced ROS 2 concepts

Once you are comfortable with the prerequisites and the concepts from this book, you can go further and learn more advanced ROS 2 concepts, such as the following:

- **Actions**: We saw actions in *Chapter 7*, but I clearly stated that this is a more advanced concept that was maybe worth skipping at the time if you felt overwhelmed. If you haven't worked on actions yet, now is a good time. Actions, along with topics and services, will allow you to use all the ROS 2 communication mechanisms between nodes.
- **Lifecycle nodes (also called managed nodes)**: These nodes contain a state machine that allows you to easily separate your code for different parts of initialization and activation. This is especially useful when dealing with hardware. For example, you can make sure that a hardware component is correctly connected and initialized before using it in a critical part of your application. Also, lifecycle nodes will be useful if you want to learn ros2_control.
- **Executors**: With executors, you can have more control over how callbacks are handled within a node or several nodes (we saw an example in the *Adding the cancel mechanism* section of *Chapter 7*).
- **Components**: By making your nodes components, you can run several nodes from within one executable. This can reduce resource usage and speed up communication. To learn about components, you first need to understand executors. Then, components will also help you understand ros2_control.

This is not a final list, but I would say that almost every ROS developer will need those concepts at some point.

On top of that, there are many additional technologies not related to ROS that can be helpful.

Extra technologies and fields

As we will see later in this chapter when we look at some job examples, being a ROS or robotics developer doesn't mean you only need to learn about ROS. There are many more things required to have a complete skill set.

Here are more technologies, fields, and tools that you can explore (I repeat, this is a non-exhaustive list, just some examples, and in no particular order):

- **Electronics/hardware**: There is a huge part of hardware in robotics. In the end, you write code to move the hardware. Without the hardware, there is nothing. This is a complete field in itself. You could learn more about hardware platforms, communication protocols, soldering components, designing a **printed circuit board** (**PCB**), and so on. Some people make a career just out of this (hardware engineers). Even if you don't plan to follow that road, it can be helpful to know a bit about it.
- **Mechanics, CAD software, 3D design**: Just like for hardware, some people make a career out of mechanical engineering. Without going too deep, it could be useful to learn about how to design a mechanical part using CAD software.
- **Fast prototyping**: This combines a lot of fields, and the goal is, as its name suggests, to create a prototype fast to validate (or not) an idea. For fast prototyping, you could use 3D printing and embedded hardware boards such as Arduino and Raspberry Pi (you can run ROS 2 on Raspberry Pi).
- **Git, continuous integration/deployment**: The chances are that you will work with other people. These tools will help you collaborate and release code more easily.
- **Docker**: The use of Docker amongst robotics developers has been rising, as you can easily set up a new environment with the correct Ubuntu/ROS 2 version. This can be extremely useful for working on several projects and testing your code in different environments.
- **DDS, networking**: ROS 2 communication relies on **Data Distribution Service** (**DDS**) and, depending on your project, you might have to dive a bit deeper into this, as well as networking in general.
- **Image processing, machine learning, and so on**: There is a huge field in robotics dedicated to analyzing the environment and extracting useful information from it. You can find a lot of ROS 2 integration with cameras, laser scanning, depth sensors, and also libraries such as OpenCV.

As you can see, that's a lot of things. When seeing this list, you might feel a bit discouraged because you realize how little you know, even after finishing a whole book about ROS 2. Don't worry, though; you don't need to learn everything. I am personally not an expert in all those fields, and nobody is. We will now see how to learn ROS 2 more efficiently by having a specific goal in mind.

Learning for a specific goal

The previous list of frameworks, stacks, and technologies can help you see the global picture of robotics, and help you pick what you need to learn next.

But, in the end, depending on your goals, your learning path may differ. If you're a student looking to finish a university project with a mobile robot, or if you need to get up to speed to work for a robotics start-up creating simulation products, the answer is going to be different.

In this section, we will explore a few examples of projects and job offers, and see what learning path is more appropriate for each example. Note that the examples we will see do not make an exhaustive list of paths. Your own path is going to be unique. The real point of this section is to show you that you should first think about what you want to do with ROS 2, and from this, pick what to learn.

Don't stress too much about it: even if you choose a path and change later, no worries. You don't need to have a definitive answer now about the kind of project or career you want to pursue. Remember, you are still in the exploration phase. Exploring means going one way, and then maybe realizing that you prefer another way, until you find something that really sticks with you. All the knowledge you gain while exploring will be valuable.

Let's dive in.

What to learn for a project?

The overall best way is to find a project and learn as you go. While building the project, you will encounter some challenges. Solving those challenges often means learning new things and they will force you to develop a better practical understanding.

Now, where to find projects? There are tons of project ideas on the internet. Depending on your hardware and financial resources, you might start with something that involves only a simulation, or if you have some hardware, you could build a robot or part of a robot yourself, and make the robot perform a task in the real world.

A few project examples

Let's consider a few projects and the learning paths they involve:

- **Using a mobile robot to find books in a library (typical university project)**: Here, you will need to make a robot navigate, so you probably have to learn about the Nav2 stack in order to map the library and make the robot move in it. On top of that, you will need to figure out how to find books, maybe with a camera. You will choose and test a camera and integrate it with your robot. This will make you practice with hardware interfacing with the robot. Then, to recognize books, you will use image processing. Also, you can start by using an existing mobile platform, and then design your own.

- **Warehouse management with mobile robots**: With this project, you will also need to make robots navigate, but this time you will have to make several robots work in the same environment in an organized way. You can learn about controlling robot swarms, which is yet another thing to learn that we haven't listed previously.

- **Maintenance patrol with a drone**: In this case, precise hardware control and remote communication will be the first challenge. Then, you will have to control the behavior of the drone. Drones are a specific kind of robot, and the MoveIt 2 or Nav2 stacks won't apply; you will need to find something else. At the time of writing, unfortunately, there is no plug-and-play stack you can use for drones in ROS 2. So, you will need to do some extra research and effort.

- **Sorting objects in a production line**: Here, we reach a completely different domain, which is the manipulation of objects. If not provided, you will need to find which robotic arm (or other device) you can use to pick and place objects, what prehension system to use, and so on. Then, you can control the robot with MoveIt 2. After that, you also need to find a way to sort products. Maybe you need a camera; in this case, you will have to correctly place the camera, do some calibration, and coordinate the whole system so that the robot knows which objects to pick, and where.

These are a few examples, but you can see that depending on what kind of robots you will use, and what you want to do with those robots, the application will be completely different, which leads to a different learning path.

My personal learning path

To give you yet another example, here is my personal story with ROS. I discovered ROS when I co-founded a robotics startup. We wanted to create an educational 6-axis robotic arm, and I was dealing with the software.

Here is, more or less, my learning path with ROS: I started with the basics, then quickly went on to create a simplified URDF for the robot, and make it move with MoveIt, so we could see it move in RViz (we didn't have a Gazebo simulation yet, but as we were building a real physical robot, we focused on the physical control first).

Then, the big challenge was to find reliable low-cost components and motors to control each of the 6 axes, create a hardware driver for each, and control them from an integrated Raspberry Pi board. This involved learning about communication protocols (and ros_control), working closely with the hardware, and experimenting a lot.

On another level, the goal was to create an intuitive user interface, so we developed some APIs on top of ROS, and a graphical interface with Angular. That alone required other kinds of skills.

The story is of course overly simplified, but as you can see, after getting the basics, I learned what I needed, when I needed it, so I could make progress on the project. For example, I didn't learn about Gazebo or the Navigation stack first; this came much later. The reason is simply because that was not the most important thing to learn at the time.

What to learn to get a job?

You might wonder: working on projects is nice, but if your goal is to get a job and start a career in robotics, what should you learn?

Again, the answer is: it depends. However, you could have a look at job offers you find on the internet and see more or less what is required for the kind of job that you want to pursue.

Also, for a junior position, it is often not expected that you are an expert in any particular field. Companies know that you have just begun, and that you still need to learn. Thus, for junior positions or internships, a strong motivation to learn (bonus point: portfolio to showcase personal projects) is often more important than having skills in x, y, or z.

Now, to give you some real examples, I found some existing job offers, and I will show you some of the technical requirements they ask for. To be clear: I won't promote any company or job offer here, I will just put a recap of technical skills they require (that I have rewritten for brevity), for example purposes only.

Job 1 – ROS 2 development internship

Here are the requirements for an internship centered around ROS 2 development:

- Python 3
- ROS 1, ROS 2
- Linux, Bash, Git
- Notions of path planning and collision avoidance
- Network and communication protocols (optional)
- C++ (optional)

To prepare for this internship, you see that you have to develop your Python, Linux, and ROS skills.

> **Note**
> They also mention ROS 1. It is worth noting that, sometimes, ROS 1 is needed (to work on legacy projects), but you can learn it on the go. Sometimes, the person writing the job description doesn't really understand the technology part of the job, and just adds all the possible keywords they encounter to make the offer look more general and attract more candidates. In this case, you will probably use ROS 2 anyway.

Path planning and collision avoidance are also included, meaning that exploring the Nav2 stack could give you a competitive advantage.

Job 2 – AI robotics internship

Here is another internship where the focus is on artificial intelligence:

- Machine learning
- Proficiency in Python, experience with PyTorch or TensorFlow
- ROS
- Enthusiasm for mobile and intelligent robots

For this internship, machine learning will probably be the most important part of the job, while working with robots that are using ROS. Thus, for a job like this, you should equally learn about machine learning, if that's your interest.

As you will see, many of the ROS jobs are not 100% centered around ROS. ROS is just one tool they use.

Job 3 – Humanoid robot development

Here are some requirements for a more intermediate job (one that requires working experience) where you participate in the development of a humanoid robot:

- ROS 2
- Linux, Git, and CI workflows
- Jetson platforms
- Robot kinematics and dynamics
- Control algorithms: force, impedance, MPC
- Communication protocols: CAN, I2C, SPI, and so on

In this job, robot control, communication protocols, and using them in an embedded Linux platform are crucial. The chances are that 80% of the challenge will be working on this, and then the rest will be about making it work with ROS 2.

I will stop here. The point was to show you that different jobs can require very different skill sets, even though they are all labeled as ROS jobs. Also, in some of them, ROS is the main focus, but many jobs use ROS only as a single tool among many others.

So, if there is a dream job you want to apply to, focus on the required skills first. To learn those skills, find a project that matches the skills, and then, as you go, find learning resources to help you complete the project.

Summary

In this chapter, we focused on what you can do after finishing this book to continue learning ROS 2 efficiently. This question can be quite tough to answer, and robotics is too vast for anyone to master everything.

Learning for a specific goal (for example, a job application or a work/school project) is the best overall since you will learn things that you can directly apply.

If you don't really know what to do, I also gave you a list of common ROS 2 stacks/frameworks, and other topics related to ROS 2 that you could learn. I strongly recommend that you learn those by doing projects, and focus on practical resources that teach you by doing.

If you liked this book and the way I teach, here are a few more resources from me:

- **Robotics Backend website** (`https://roboticsbackend.com/`): Here, you will find more written tutorials about ROS 2 and other robotics-related topics
- **Robotics Backend YouTube channel** (`https://www.youtube.com/c/RoboticsBackend`): For video tutorials and free crash courses
- **Full-length online courses** (`https://roboticsbackend.com/courses/`): I also provide complete ROS 2 courses that you can purchase, with a strong focus on practical learning

As we saw in this chapter, you have now finished the discovery phase, and you enter the exploration phase. As you make progress and get better at what you do, you will start to find which field in particular you want to dive into. Depending on this, and the opportunities you get, you will start to specialize.

Until then, don't overthink anything, and just learn/explore as much as possible. Start several projects, learn other technologies, and be curious. Also, don't forget to have fun while learning and building projects. This is one of the most important things that will motivate you to go further.

I wish you good luck on your journey and hope this book has helped you get started on the right foot!

Index

Symbols

2D robot simulation
　running 36-38
_bringup package 295

A

action 48-50
　client, writing 168
　goal, sending from terminal 49
　interface (data type) 48, 49
　introspecting 181, 182
　listing 181, 182
　name 48
　services 183, 184
　topics 183, 184
action mechanisms 173
action server
　running 48
　writing 161
actuator 8
AI robotics internship 345

C

C++ 11, 340
　parameters 194
C++ action client
　writing 171, 172
C++ action server
　callbacks, implementing 166, 167
　setting up 165, 166
　writing 165
C++ node
　building 77
　creating 73
　running 77
　template 79
　writing 74-76
C++ parameter callback 204
C++ publisher
　building 95, 96
　running 95, 96
　used, for creating node 94, 95
　writing 94
C++ service client
　writing 138, 139

C++ service server
  service interface, importing 132
  service server, adding in node 133, 134
  writing 132
C++ subscriber
  running 100
  used, for creating C++ node 99, 100
  writing 99
callback_pose() method 145
call_set_pen() method 145, 207
cancel mechanism 156, 157
  adding 176
  problem 176, 177
  with C++ 180, 181
  with Python 177-179
client node 43
client side service
  solution 144-146
closed loop control 115
  challenge 115
  challenge, solution 116, 117
CMake extension 30
collision element
  defining 315
collision tags 315
collision tags, URDF for Gazebo
  collision element, defining 315, 316
  collision properties, adding
    for links 316, 317
  collision properties, validating
    with RViz 317, 318
Computer-Aided Design (CAD) 255
controller 8
create_timer() method 72
custom action interface
  creating 157, 158
custom interface
  challenge 143
  creating, for ROS topic 107
  existing interfaces, using 108
  topic interface, creating 110, 111
custom meshes
  installing 294
custom service interface
  creating 124-128
  existing interface, finding for
    service 124, 125

D

Data Distribution Service (DDS) 101, 341
data types
  Booleans 191
  integer array 191
  string 191
declare_parameter() method
  used, for using parameters 191
dependency loop 212
development tools, ROS 2
  terminator 30, 31
  Visual Studio Code (VS Code) 29
differential drive controller 326
differential drive (diff drive) 326
discovery phase 336
distributions 13

E

end-of-life (EOL) 15
existing interfaces
  existing messages, avoiding 109, 110
  finding 108
  using 108
  using, in code 109
exploration phase 336

F

feedback mechanism 155-157
 adding 173
 with C++ 174, 175
 with Python 173, 174
frames per second (FPS) 188
frameworks 336

G

Gazebo 6, 305-308
 used, for controlling robot 323
 used, for spawning robot 318
 versus RViz 304, 305
 working 304
 working, with ROS 2 308, 309
Gazebo and ROS 2 communications
 bridging 328, 329
 configuration file, adding to bridge topics 329, 330
 Gazebo bridge, with YAML configuration file 330, 331
Gazebo Harmonic 306
Gazebo system
 differential drive controller 326, 327
 joint state publisher 328
 Xacro file 325, 326
goal
 sending, from terminal 182, 183
graphical user interface (GUI) 288

H

hardware interface 337
humanoid robot development 345

I

inertial tags, URDF for Gazebo
 inertial macros, adding for basic shapes 311, 312
 inertial macros, adding in links 312-314
 inertial macros, writing 310, 311
 inertial property, validating with RViz 314
integrated development environment (IDE) 29
interface 40, 85
interface, action
 feedback 48
 goal 48
 result 48

J

job 344
 AI robotics internship 345
 humanoid robot development 345
 ROS 2 development internship 344
joint attribute
 name 262
 type 262
joints 253
joint state publisher 286, 328
joint tags
 <child> tag 263
 <origin> tag 263
 <parent> tag 263
joint type
 reference link 266

L

launch file 52-54
 _bringup package, creating 321
 starting 52, 53

used, for spawning robot 320
writing 322
writing, to publish TFs 295
writing, to visualize robot 295
links and joints
  assembling process 261
  joint, adding 262-264
  joint origin, fixing 264, 265
  joint type, setting up 265, 266
  process 267, 268
  second link, adding 261, 262
  visual origin, fixing 266, 267
Linux 339
Linux command line 11
Long-Term Support (LTS) 15

M

mobile robot
  base footprint 274, 275
  caster wheel, adding 273, 274
  final result 269
  used, for writing URDF 268
  wheels, adding 269
mobile robot, wheels
  left wheel link 272, 273
  right wheel link 270-272
Moveit 2 338, 339
my_robot_bringup package 295

N

name 85
namespace 225
  adding, in XML launch file 227
  node, adding 225, 226
Navigation 2 stack (Nav2) 337

new action interface
  creating 158-160
node 6, 33, 38, 39
  2D robot simulation, running 36-38
  introspecting, with rqt_graph 35, 36
  running 34
  starting, from terminal with ros2 run tool 34
node introspection 80
  node name, changing at run time 81
  ros2 node command line 80, 81
node template
  for C++ node 78, 79
  for Python node 78, 79
number_publisher node
  parameters, using 190

O

Object Oriented Programming (OOP) 11, 66
odometry (odom) 327
operating system (OS) 13

P

parameter 50-52, 187
  additional tools, handling 198
  at runtime 192-194
  challenge 205
  data types, recapping 197
  declaring, with Python 191, 192
  exporting, into YAML 199
  for multiple nodes 196
  loading, from YAML files 195
  obtaining, for node 50, 51
  obtaining, with Python 191, 192
  services 201, 202
  solution 206, 207

Index   351

storing, in YAML files 195
updating, with parameter callback 202, 203
using, in number_publisher node 190
value, setting from terminal 200, 201
values, getting from terminal 198
value, setting up for node 51
with C++ 194
with Python 191, 192
parameter callback 187, 202
used, for updating parameters 202, 203
plumbing 6
printed circuit board (PCB) 341
process ID (PID) 215, 307
project
examples 342
learning 342
personal learning path 343
property 276
publisher 39, 84
Python 11, 339
and C++ nodes, running together 101
parameters 191, 192
used, for declaring parameters 191, 192
used, for obtaining parameters 191, 192
Python action client
callbacks, implementing 169, 170
communication, trying 171
creating 168, 169
writing 168
Python action server
goal, accepting 163
goal, executing 163, 164
goal, rejecting 163
setting up 161, 162
writing 161
Python launch file 298-300
adding, in XML launch file 218, 219
challenge 228, 229

combining with 218
communications, renaming 220
creating 215
issue 217, 218
nodes, configuring 219-221
nodes, renaming 220, 221
parameters, setting 221
parameters value, adding 222
solution 229-232
versus XML launch file 217
writing 216, 217
YAML param files, installing 223, 224
YAML param files, loading 223, 224
Python node
building 69, 70
callback function 71
creating 66
file, creating 67
improving 71-73
minimal ROS 2 Python node, writing 67-69
running 71
template 78, 79
timer, adding 71
Python parameter callback 203, 204
Python programming 11
Python publisher 89
adding, to node 90, 91
building 92, 93
node, creating 89
publishing, with timer 91
running 93, 94
Python service client
writing 135-137
Python service server
request validating 131, 132
service interface, importing 128, 129
service server, adding to node 129, 130
writing 128

Index

Python subscriber
   running 98
   used, for creating Python node 97, 98
   writing 97

R

red, green, blue (RGB) 51
request 44
response 44
robot
   spawning, from launch file 320
   spawning, from terminal command 319, 320
   spawning, in Gazebo 318
robot control
   Gazebo system, adding 325
   in Gazebo 323
   testing 331, 332
   TF errors, in RViz 324
Robotics Stack Exchange 7
robot model
   visualizing, in RViz 238-240, 288
Robot Operating System (ROS) 3-5
   capabilities 7
   examples 8, 9
   framework with plumbing 6
   need for 4, 5
   online community 7
   set of tools 6
   usage scenario 8
robot_state_publisher node 284
robot_state_publisher node, inputs
   joint states topic 285
   URDF as parameter 285
ROS 1
   hardware prerequisites 11
   knowledge prerequisites 11
   prerequisites 10

   software prerequisites 11
   versus ROS 2 9, 10
ROS 2
   development tools 29
   used, for working with Gazebo 308, 309
ROS 2 actions
   example 152, 153
   working 154-156
ros2 command-line tool 35, 139
   service name at runtime, changing 141
   service request sending 141
   services, introspecting 140
   services, listing 140
ROS 2 communication
   services 152
   topics 152
ROS 2 development internship 344
ROS 2 distribution 14, 15
   LTS distribution 15
   non-LTS distribution 15
   reference link 14
   selecting 14
ROS 2 environment
   setting up 27
   source line, adding to .bashrc file 28
   sourcing 28
ROS 2 functionalities
   service callback 130
   service interface 130
   service name 130
ROS 2 Humble 10
ROS 2 installation, on Ubuntu 25
   locale, setting 25
   ROS 2 packages, installing 26, 27
   sources, setting up 25, 26
ROS 2 launch file 210
   example, with seven nodes 211
   need for 210, 211

ros2 node command line 80, 81
ROS 2 nodes 123, 124
ROS 2 package 61
  building 64, 65
  C++ package, creating 63, 64
  creating 61
  nodes, organizing 65, 66
  Python package, creating 62, 63
  ROS-Base 26
  ROS Desktop 26
ROS 2 parameter 188-190
  need for 188
  node example 188, 189
ROS 2 roadmap 336-339
  advanced concepts 340
  basics 339
  discovery phase 336
  exploration phase 336
  specialization phase 336
  technologies and fields 340, 341
ros2 run tool
  used, for starting node from terminal 34
ROS 2 services 120
  client 120
  multiple clients for one service 121
  robotics example 122
  server 120
ROS 2 topic 84
  command line 103, 104
  custom interface, creating 108
  data, replaying with bags 106, 107
  handling, tools 101
  introspecting, with rqt_graph 102, 103
  multiple publishers and subscribers 86
  multiple publishers and subscribers, inside one node 86-88
  name, changing at runtime 105, 106
  publisher and subscriber 84

  using 88
  working 88
ROS 2 workspace 58
  building 59
  creating 58, 59
  sourcing 60
ROS actions
  working 156, 157
ROS Discourse forums 7
ROS distribution
  selecting 16
ROS Jazzy 14
  reference link 17
ROS Jazzy Jalisco 14
ROS Rolling 15
rqt_graph
  used, for introspecting nodes 35, 36
  used, for introspecting ROS 102, 103
RViz 237
  configuration, installing 295
  configuration, saving 290
  configuring 288, 289
  installation 238
  setting up 238
  used, for visualizing robot model 238, 288
  versus Gazebo 304, 305

S

sensor 8
server 43
server side service
  solution 146, 147
service client 120
  challenge 143
  client and server node, running 137, 138
  writing 135

services 43, 47
  challenges 45
  challenges, solution 46, 47
  experimenting with 45
  interface (data type) 44, 45
  name 44, 45
  request, sending from terminal 45
  server, running 43, 44
service server 120
  challenge 143
  writing 128
Simultaneous Localization And Mapping (SLAM) 338
specialization phase 336
stacks 336
subscriber 39, 84

T

terminal
  command, used, for spawning robot 319, 320
  nodes 287
  robot model, visualizing in RViz 288
  TFs, publishing 287, 288
terminator 30, 31
topic interface
  building 111-113
  creating 110-113
  custom message, using in code 113, 114
  package, creating 110
  package, setting up 110, 111
topic publisher
  C++ publisher, writing 94
  Python publisher, writing 89
  writing 88

topics 39, 43
  challenges 41
  challenges, solutions 41, 42
  experimenting with 41
  interface (data type) 40
  name 40
  publisher, running 39
  subscriber, running 39
topic subscriber
  C++ subscriber, writing 99-101
  Python subscriber, writing 97, 98
  writing 96
TransForms (TFs) 237, 240-243
  3D coordinate frame, tracking 249
  computing 250
  links 241
  parent and child relationship 244, 245
  problem solving 249
  publishing 286
  publishing, with URDF 284
  relationship between 244
  robot_state_publisher node 284, 285
  robot_state_publisher node inputs 285
  /tf topic 245-247
  TF tree, visualizing 247, 248
TurtleBot 14
Turtlesim 14

U

Ubuntu 11
  installing 16
  relationship, with ROS 2 17
Ubuntu 24.04 17
  installing, with dual boot 17, 18
Ubuntu 24.04 installation, on VM 18
  finishing 22, 23
  Guest Additions CD Image 23, 24

new virtual machine, creating 19-22
starting 22, 23
Ubuntu .iso file, downloading 18
VirtualBox, installing 18
Unified Robot Description Format (URDF) 239, 253
  creating, with link 254
  _description package, creating 292, 293
  file compatible, making with Xacro 276
  file, installing 293
  file, setting up 254
  files, installing 293, 294
  improving, with Xacro 276
  installing, to create package 291
  link, creating 255
  ros2_ws workspace, adding 291, 292
  used, for publishing TFs 284
  visualizing, in RViz 256, 257
  visual link, modifying 258
  visual origin, modifying 257
  writing, for mobile robot 268
  XML code link 255, 256
unused parameter 175
URDF for Gazebo 309
  collision tags 315
  inertial tags 310
urdf_tutorial package 239

V

virtual machine (VM) 13, 304
visual link
  color, defining 259, 260
  modifying 258
  shapes, defining 258, 259
Visual Studio Code (VS Code) 29

W

workspace 58

X

Xacro 253
  used, for improving Unified Robot Description Format (URDF) 276
Xacro file
  adding, in another file 280, 281
  common_properties.xacro file 280
  files, installing 293, 294
  mobile_base.xacro file 280
  my_robot.urdf.xacro file 280
Xacro macros 279, 280
Xacro properties 277, 278
XML launch file 295
  challenge 228, 229
  combining with 218
  communications, renaming 220
  creating 212
  installing 212-215
  launch file, writing 296, 297
  launch folder, creating 296
  launch folder, installing 296
  launch folder, starting 298
  nodes, configuring 219-221
  nodes, renaming 220, 221
  package, setting up 212, 213
  parameters, setting 221
  parameters value, adding 222
  solution 229-232
  starting 214, 215
  versus Python launch file 217
  writing 213, 214
  YAML param files, installing 223, 224
  YAML param files, loading 223, 224

Y

YAML files 187
  parameters, loading 195
  parameters, storing 195
YAML launch files 222
YAML param files 222

packtpub.com

Subscribe to our online digital library for full access to over 7,000 books and videos, as well as industry leading tools to help you plan your personal development and advance your career. For more information, please visit our website.

Why subscribe?

- Spend less time learning and more time coding with practical eBooks and Videos from over 4,000 industry professionals
- Improve your learning with Skill Plans built especially for you
- Get a free eBook or video every month
- Fully searchable for easy access to vital information
- Copy and paste, print, and bookmark content

Did you know that Packt offers eBook versions of every book published, with PDF and ePub files available? You can upgrade to the eBook version at packtpub.com and as a print book customer, you are entitled to a discount on the eBook copy. Get in touch with us at customercare@packtpub.com for more details.

At www.packtpub.com, you can also read a collection of free technical articles, sign up for a range of free newsletters, and receive exclusive discounts and offers on Packt books and eBooks.

Other Books You May Enjoy

If you enjoyed this book, you may be interested in these other books by Packt:

Artificial Intelligence for Robotics

Francis X. Govers III

ISBN: 978-1-80512-959-2

- Get started with robotics and AI essentials
- Understand path planning, decision trees, and search algorithms to enhance your robot
- Explore object recognition using neural networks and supervised learning techniques
- Employ genetic algorithms to enable your robot arm to manipulate objects
- Teach your robot to listen using Natural Language Processing through an expert system
- Program your robot in how to avoid obstacles and retrieve objects with machine learning and computer vision
- Apply simulation techniques to give your robot an artificial personality

Hands-on ESP32 with Arduino IDE

Asim Zulfiqar

ISBN: 978-1-83763-803-1

- Understand the architecture of ESP32 including all its ins and outs
- Get to grips with writing code for ESP32 using Arduino IDE 2.0
- Interface sensors with ESP32, focusing on the science behind it
- Familiarize yourself with the architecture of various IoT network protocols in-depth
- Gain an understanding of the network protocols involved in IoT device communication
- Evaluate and select the ideal data-based IoT protocol for your project or application
- Apply IoT principles to real-world projects using Arduino IDE 2.0

Packt is searching for authors like you

If you're interested in becoming an author for Packt, please visit `authors.packtpub.com` and apply today. We have worked with thousands of developers and tech professionals, just like you, to help them share their insight with the global tech community. You can make a general application, apply for a specific hot topic that we are recruiting an author for, or submit your own idea.

Share Your Thoughts

Now you've finished *ROS 2 from Scratch*, we'd love to hear your thoughts! Scan the QR code below to go straight to the Amazon review page for this book and share your feedback or leave a review on the site that you purchased it from.

`https://packt.link/r/1835881416`

Your review is important to us and the tech community and will help us make sure we're delivering excellent quality content.

Download a free PDF copy of this book

Thanks for purchasing this book!

Do you like to read on the go but are unable to carry your print books everywhere?

Is your eBook purchase not compatible with the device of your choice?

Don't worry, now with every Packt book you get a DRM-free PDF version of that book at no cost.

Read anywhere, any place, on any device. Search, copy, and paste code from your favorite technical books directly into your application.

The perks don't stop there, you can get exclusive access to discounts, newsletters, and great free content in your inbox daily

Follow these simple steps to get the benefits:

1. Scan the QR code or visit the link below

https://packt.link/free-ebook/9781835881408

2. Submit your proof of purchase
3. That's it! We'll send your free PDF and other benefits to your email directly

Made in United States
Troutdale, OR
03/17/2025